BRITISH ENGINEERS AND AFRICA, 1875–1914

Empires in Perspective

Series Editors: *Tony Ballantyne*
                     Duncan Bell
                     Francisco Bethencourt
                     Caroline Elkins
                     Durba Ghosh
Advisory Editor: *Masaie Matsumura*

Titles in this Series

FORTHCOMING TITLES

BRITISH ENGINEERS AND AFRICA, 1875–1914

BY

Casper Andersen

Routledge
Taylor & Francis Group

LONDON AND NEW YORK

First published 2011 by Pickering & Chatto (Publishers) Limited

Published 2016 by Routledge
2 Park Square, Milton Park, Abingdon, Oxfordshire OX14 4RN
711 Third Avenue, New York, NY 10017, USA

First issued in paperback 2015

Routledge is an imprint of the Taylor & Francis Group, an informa business

© Taylor & Francis 2011
© Casper Andersen 2011

BRITISH LIBRARY CATALOGUING IN PUBLICATION DATA

Andersen, Casper.
British engineers and Africa, 1875–1914. – (Empires in perspective)
1. Engineering – Great Britain – History – 19th century. 2. Engineering – Great Britain – History – 20th century. 3. Great Britain – Colonies – Africa – History – 19th century. 4. Great Britain – Colonies – Africa – History – 20th century. 5. Great Britain – Foreign relations – Africa. 6. Africa – Foreign relations – Great Britain.
I. Title II. Series
620'.00941'09034-dc22

ISBN-13: 978-1-138-66148-6 (pbk)
ISBN-13: 978-1-8489-3118-3 (hbk)

Typeset by Pickering & Chatto (Publishers) Limited

CONTENTS

For Ulla and Loa and for my parents Finn Andersen and Inger Andersen

ACKNOWLEDGEMENTS

The research for this book has been funded by a scholarship from the Faculty of Humanities, University of Aarhus, and by a postdoctoral grant from Carlsberg Fondet (grant number 2009_01_0676). The project received additional funding from Knud Højgaards Fond, Oticon-Fondet, and the Danish Research School for Philosophy, History of Ideas and History of Science. I wish to thank these funding bodies for the financial support that made it possible to conduct the research for this book.

In writing this book I have been fortunate to benefit from the help of colleagues, friends and family. There are some whom I would particularly like to mention. At the University of Aarhus my colleague, friend and former supervisor Peter C. Kjærgaard has been a great support and over the years he has taught me what it takes to be a historian. I would also like to thank Mikkel Thorup for his motivating encouragement and solid advice. Henry Nielsen, whose inspiring teaching and writing first aroused what at the time seemed an unlikely interest in engineers and their history, has been a great help in sharpening my take on imperial engineers and it has been a pleasure for me to discuss this project with him. Several people in Aarhus have taken time to read, comment on and discuss draft chapters and their useful suggestions have helped me to improve the arguments substantially. In this respect I would like to thank especially Henning Høgh Laursen, Mats Fridlund, Stine Grumsen and Patrick Luke Cockburn.

Much of the research for this book has been carried out in Britain, a country whose people and history have become still more important to me. I am grateful to a number of people and institutions there. The professionalism, helpfulness and, indeed, kindness of the entire staff at Rhodes House Archive, Oxford, has made it a true pleasure to spend long days of research in that building. At Radcliffe Science Library in Oxford, where I was in the habit of requesting heavy, dusty engineering volumes undoubtedly kept in the most remote corner of the deposits, the librarians have also been very helpful. I have also benefited from the help of the staff at the library and archive of the Institution of Civil Engineers in London where Carol Morgan has guided me to valuable sources and where Mike Chrimes has shared generously of his profound knowledge of the history of the

ICE and the British civil engineering profession in general. At the Egypt Exploration Society, Patricia Spencer has offered excellent assistance when I conducted research in the society's fascinating archive. I am also indebted to the people at the Department for the History and Philosophy of Science at the University of Leeds where I spent five inspiring months in the autumn of 2006. I would in particular like to thank Graeme Gooday for commenting on early, untidy papers of mine, for hosting more than one good party and, in particular, for sharing generously of his unsurpassed knowledge of Victorian engineers and their histories. Several scholars have found the time to discuss my ideas during my research stays in Britain. They include Jon Topham, Diane Drummond, Andrew Thompson, Andrew Cohen, Peter Cain, Christine Macleod, Eileen Magnello, Simon Schaffer, Malcom Dunkeld, Ian Phimister and in particular David Sunderland. Their advice pushed me in very different but always very useful directions. I owe a special word of thanks to Jan-Georg Deutsch at St Cross College in Oxford, where I spent five wonderful months as a visiting academic in the spring of 2008. Georg has been tremendously supportive of me and my work for the last six years. For his friendship and support I am very grateful. Ben Marsden and John Darwin provided critical and encouraging comments to the project and without their expert suggestions, criticism and ideas this book would have been a lot poorer. I am indebted also to Ian Kerr for recommending the manuscript to Pickering & Chatto. Two anonymous readers appointed by Pickering & Chatto assessed the book proposal and manuscript and I am grateful for their comments, which helped me improve the manuscript considerably. At Pickering & Chatto I would also like to thank the commissioning editor, Daire Carr, and the editorial manager, Julie Wilson, for their assistance during the publication process.

I have several people to thank on a more personal note. Hans Henrik Hjermitslev has commented engagingly on drafts of the entire manuscript and has been a great moral support when dead engineers have been less willing to collaborate with the historian. A warm thanks to Bob for inspiring rock and lyrics and to Claus Andersen who assisted me in putting the final letters to the page and who prior to this has helped me more times than I can count. Finally, a very special word of gratitude to my partner and favourite companion Ulla Hjorth Jørgensen for her unwavering encouragement, patience and support and to Loa Hjorth Andersen who recently joined our journey and made it even more joyful than it was before.

LIST OF TABLES

INTRODUCTION

In the British empire engineers were important agents, and this is a book about them. More specifically it is a study of the imperial diasporas, identities and networks that developed as the British engineering profession established connections on the African continent in the period 1875–1914. The book combines and integrates in new ways perspectives from the fields of imperial history and history of science and technology with the purpose of analysing the imperial connections of the engineers. The methodological approaches employed in the six chapters of the book are introduced in the opening sections of the individual chapters, while it is the purpose of this introduction to establish the overarching historiographical frame of the book and to explain what the main issues that it addresses are.

The Empire at Home and its Connections

The last quarter of the nineteenth century was a period in which the boundaries of the British empire widened drastically, including in the African continent, which was divided in a scramble spurred by economic and strategic rivalries among the European powers.[1] Technologies in the form of medicine, improved military equipment and infrastructural systems such as harbours facilities, railways and telegraphs were important factors in this process.[2] These infrastructural technologies provided the muscle for conquest and were actively employed in strategies for formal empire-building. Moreover, they also served as a 'main generator of those insidious partnerships of imperial, financial, and commercial interests that go into the making of "informal empires"'.[3]

Hence, this was a period in which imperialism and engineering became closely intertwined. In the second half of the nineteenth century, as Robert Kubicek notes, 'civil engineers abroad played a significant if insufficiently emphasized role in transferring infrastructure and disturbing the status quo'.[4] Indeed, over this period Britain's industrial pre-eminence ultimately growing out of the Industrial Revolution 'instigated a swarming of engineers to foreign and colonial adventures', creating what Robert Angus Buchanan has labelled 'The Diaspora of British engineering': after the 1830s, British engineers moved in still wider geographical circles and in the process left an enduring stamp across the globe, in particular in the form of bridges, dams, harbours and railways. British engineers who '(1) recognised that they possessed an expertise that was in short supply elsewhere and (2) were prepared to travel abroad in large numbers in order to provide it' first expanded their operations to Continental Europe and North America, and after 1850 also began to work in Latin America, Australia and Asia. Eventually they made their way into Africa.[5] Between 1875 and 1914 British engineers designed harbours along the coast of Africa, scaled the Rift Valley with dozens of steel viaducts, dammed the Nile, and laid out railway tracks so that the length of British-built lines in Africa in 1907 had exceeded 10,000 miles, more than half of the total combined length of railways in Africa at the time.[6]

This part of the history is in many respects familiar. Its tangible legacy is still visible; the bridge across the Zambezi River by the Victoria Falls and the first and now smaller Aswan Dam, to mention two projects discussed in this book. The issues also, at first sight, fit comfortably within a familiar, slightly dated, perhaps eclipsed, historiography that has explored how big, masculine, industrial technologies were diffused from Europe and impacted on the rest of the world with little 'local' input – diffused either as a boon bestowed by advanced civilizations upon 'backward' societies or as oppressive tools of empire that enabled European penetration, conquest and consolidation across the globe.[7] Disagreements about causes and consequences are deep and important within this historiographical framework but it is based on the underlying assumption that what needs explaining is how 'Western', and in this case British, engineering impacted on the colonial world during 'the age of empire'.[8]

During the last decades historians have forcefully challenged and rightly dismissed 'diffusionist' approaches to the study of technology in the context of imperialism, emphasizing in particular the inability of the perspective to account for developments in the colonial world. This book also goes in different directions from the 'diffusionist' historiography. A first step is to reverse the question and examine not only how colonial regions were influenced by British engineering, but also how the civil engineering profession in Britain was influenced by experiences of imperialism in this period. The outward movement of British engineers impacted on developments throughout Britain's formal and informal empire but it had equally strong and often more enduring repercussions in Britain. Indeed, it is a central claim of this book that the outward movement of British civil engineering strongly influenced the profession and its place in British society and culture. It argues that key developments in the civil engineering profession need to be analysed against wider geographical and ideological contexts that in this period were closely tied to British imperialism. As will be demonstrated, 'imperial factors' were evident and of growing importance in the central institutions of the profession, in the business platforms that leading civil engineers operated from, in the public perception of civil engineers and in relation to more slippery notions concerning the identity of British civil engineers and the ideological outlook of the profession.

It is from the field of imperial history that this book adopts the insight that empire and imperialism were not only something 'out there' but also factors that strongly affected developments in British society. Scholars have long insisted on the importance of empire in relation to British society, culture and national identity and have from a range of perspectives, analysed its impact on Britain in politics, popular and elite culture, feminism, science and a range of other contexts.[9] This has sparked a prolonged debate about whether British society and culture were 'saturated with empire' or if other factors, notably class, were more important in Britain even in the high age of imperialism.[10] In a useful intervention Andrew Thompson has insisted that it is unfruitful to discuss the 'amount' of imperialism in Britain because this fails to capture what diverse and pluralistic entities both Britain and the empire were. He rightly asserts that

> Whether they were explorers, traders, settlers, soldiers, missionaries or officials, the people of Britain became caught up in the process of overseas expansion not only in vastly different but unequal ways. There was never likely to be any single or monolithic 'imperial culture' in Britain, therefore.[11]

Thompson convincingly demonstrates that the empire 'struck back' into British society in unequal ways for different classes, genders, regions and indeed for different professional groups. This book follows this line of inquiry by exploring the ways in which empire 'struck back' at the engineering profession while knowing that it was more likely to do so in complex, unequal ways.

For this line of enquiry it is imperative to distinguish between different kinds of imperial influences that affected the profession. An important part of this influence came from the substantial groups of British engineers who were based in the colonial world but who retained strong connections with the engineering communities in their mother country. In 1902, 20.6 per cent of the 6,414 members of the Institution of Civil Engineers (ICE) in London were resident in British colonies (see Table 4.1 below). The ICE was the most prominent of the accredited institutions of the British engineering profession. The position of the large minority of 'colonial' members in the ICE is analysed in detail in Chapter 4 and the book more generally investigates to what extent, and through what channels, the influence of these large expatriate communities fed back into the British engineering profession.

Imperial impulses did not only originate from engineers in 'colonial diasporas'. Indeed, the primary focus of this book is with London-based consulting engineers whose imperial platform was metropolitan rather than colonial. Operating from offices in Westminster, this powerful segment of the engineering profession based their income and professional status on planning and designing large infrastructural projects overseas and in particular in the regions of the world where Britain's formal and informal imperial power was paramount.

The imperial influences often stemmed from, and even more frequently filtered through, this elite segment of the profession that occupied a position enabling it to affect the engineering profession far beyond their own closed ranks and tightly knit networks. Furthermore, between the expatriate communities in the colonies and the metropolitan consulting engineers, many British civil engineers spent their lives and careers moving back and forth between Britain and the engineering frontiers of the empire (or occasionally from one engineering frontier to the next) in search of fortune, fame or simply the next job opening. Such engineers and their complex imperial diasporas and careers are important because they enable us to flesh out connections between London and shifting regions overseas, connections that otherwise can be difficult to pinpoint.

In the British Isles there were large concentrations of engineers in locations apart from London. Glasgow is a notable case in point from where imperial connections also developed with Africa in this period.[12] The focus in this book is, however, on London, which during the period constituted a vital imperial engineering hub. By reconstructing and analysing the diverse engineering connections and impulses in London an image emerges of a dynamic metropolitan engineering centre; never self-contained, always connected and constantly reconfiguring connections with a wider imperial world. This view resonates with recent re-conceptualizations of the British empire as a zone bound together by 'colonial connections' and 'imperial networks' through which knowledge circulated, people travelled, and trust and authority were negotiated.[13]

Thus, the purpose of reversing the perspective to include Britain and to devote substantial attention to London-based engineers is not to argue in favour of a Eurocentric model in which the causes of a technologically driven globalization are seen as originating from 'the imperial centre' and from there diffused to more or less passive 'colonial peripheries'. Nor is it the purpose to engage in the debates on whether the forces of Britain's imperial expansion were primarily centrifugal and metropolitan or centripetal and peripheral.[14] Rather, the point is that it brings into focus the mutually constitutive connections between Britain and Africa during this period. It is such connections and in particular their reciprocal nature that this book is particularly concerned with. In order to explore these connections, analytical frames are required that are capable of encompassing Britain as well as colonial regions. To pursue this inquiry with respect to engineers, a useful concept is that of 'bridgeheads'. The term was introduced in the historiography by John Darwin, who defined the bridgehead as 'the hinge or "interface" between the metropole and a local periphery' and he emphasized the diversity of bridgeheads that might, for example, 'be a commercial, settler, missionary or proconsular presence or a combination of all four' and might consist of 'a decaying factory on a torrid coast or, at its grandest, the "Company Bahadur"'.[15]

Thinking in terms of bridgeheads has great potential. As Alan Lester has rightly asserted, the concept recognizes the pluralism of British society and 'the co-existence of different British interests, each with their own ways of connect-ing metropole and colony'.[16] It thereby draws attention to the fact that there were 'multiple, and often contestatory 'projects' of colonialism'.[17] Moreover, as Lester notes, 'conceiving of several "bridgeheads" connecting any one colony with Britain, Darwin was close to elaborating a networked or webbed concep-tion of imperial space also characteristic of the "new" imperial history'.[18] These concerns are central to this book. The insistence on multiple points of contact is prerequisite for exploring the dynamics between members of the engineer-ing profession and other groups with imperial interests in Britain and Africa. The chapters that follow also bring into focus contestatory imperial 'projects' among engineers as well as tensions with the 'projects' of other imperial groups. The book, moreover, reshuffles spatial categories in order to make sense of the multilayered engagements between the British engineering profession and forms of imperialism during this period. It demonstrates that whatever is left of the schism between 'traditional' and 'new' imperial history must be straddled in order write a history of the engineers that takes into account the economic and political as well as the cultural and ideological dimensions of the imperial con-nections they developed.[19]

Consulting Engineers and the Great George Street Clique

An important awareness that historians of the British empire can learn from his-torians of engineering is that the category 'engineer' was multifaceted, complex and subject to change over time.[20] In the British context an important distinc-tion was that between Royal Engineers and civil engineers, the latter initially meaning that minority of engineers who were not in military service.[21] During the nineteenth century specialization and growth in the engineering profes-sion, however, meant that terms such as mechanical engineer, marine engineer, mining engineer and later electrical engineer became common ways of distin-guishing between different kinds of non-military engineers and that the term 'civil engineer' gradually came to mean a specific kind of non-military engineer that was engaged, in particular, in infrastructural engineering. However, among civil engineers and in particular in the ICE there was a strong adherence to the idea that the 'civil engineer' was more than a subcategory among other categories and that it retained at least part of its overarching meaning: civil engineering was the stem of the engineering tree from which the other forms had branched out and which therefore remained indebted to civil engineering. Naturally, engineers and professional institutions in fields such as mechanical or electrical engineering did not always share this point of view.[22]

A further complication in this period – and the one most important for this book – related to the category of the consulting engineer.[23] The term 'consulting engineer' was a contested and ambiguous category, an issue that contemporaries also struggled with. In 1909 the London-based lawyer W. Valentine Ball published a book on *The Law Affecting Engineers* in which he noted:

> We have the terms 'civil engineer'; 'mechanical engineer'; 'electrical engineer'; 'mining engineer'; 'marine engineer'; and 'railway engineer'. Last, but not least, we are accustomed to hear the phrase 'consulting engineer'. In what particular branch of the profession he holds himself out for consultation does not always appear. Nor is it necessary that he shall have any particular experience in the art of consulting or advising in consultation. Any man can assume this imposing title; and there is no disciplinary body to whom persons who consult him can complain if they find out that he has no qualification either as an engineer or as a consultant.[24]

Indeed, there was no institutional body specifically devoted to consulting engineers until the very end of the period studied in this book.[25] However, in spite of its judicial vagueness and lack of institutional back-up the term 'consulting engineer' was ubiquitous among engineers at the time and as Ball also pointed out 'among the leaders of the profession it would seem that the phrase "consulting engineer" has a well-defined meaning'.[26] This 'meaning' will become clearer in the course of the book but central elements can be summarized at the outset. The term was used, primarily, to denote independently practising engineers. In this period most engineers were employed by state or private companies for example in mining, manufacturing or railways, but the consulting engineers were the minority that could claim the elevated status of 'independent professionals'.[27] Consulting engineers were designers rather than builders. They were hired by promoters of engineering projects to carry out surveys, to estimate cost, prepare designs, draw up specifications, inspect purchased equipment and supervise construction processes. The consulting engineer to a project retained responsibility for the overall design of the works while contractors were responsible for recruiting and organizing labour, for supplying the workers with tools, for setting them to work and for paying their wages.[28] Moreover, the consultants received fees for their services and were not allowed to take on any contracts. The idea was that the consulting engineer should be independent of contractor and manufacturing interests so that the loyalty of the consultant would be undivided and directed only towards providing the best engineering solutions to the challenges of a project.[29] This ideal of 'independence' was crucial; the key value, essential character trait and defining marker in the professional and social identity of consulting engineers. It was an ideal, moreover, that was in accord with the gentlemanly aspirations of the top layer of consulting engineers. Indeed, the term 'consulting engineer' referred to certain professional functions but also to the status of an engineer in complex socio-professional hierarchies that existed among British civil engineers – hierarchies that by the last quarter of the nineteenth century consisted of pupils, clerks, resident, assistant and chief engineers, with the independent consulting engineers at the top.[30]

In the second half of the nineteenth century, the civil engineering consultancy business was dominated by a network of professionals based in central London. In an important article Dale Porter and Gloria Clifton referred to these consulting engineers as the 'Great George Street Clique', and this is very accurate.[31] Their offices in Westminster clustered around this street where the ICE was located also. The consulting engineers of the Great George Street Clique dominated the ICE as well as the market for engineering consulting to projects in the British empire. An immediate impression of these facts may be obtained from Table 3.1 below. They formed a diverse group but as a whole the Westminster consulting engineers were affluent, vocal, influential and self-confident. It was, furthermore, a group whose status and position was strengthened as the consultants became occupied increasingly with designing engineering projects throughout the empire.

From a privileged position in Westminster the men (and they were indeed all men) of the Great George Street Clique made connections with shifting regions in Britain's formal and informal empire and after 1875 also with those in Africa. This intermediary function was a seminal characteristic of the consulting engineers of the Great George Street Clique and they can be seen as paradigmatic examples of what John Darwin has referred to as the 'second bridgehead' located 'at the domestic end of the imperial axis' and consisting of enclaves of 'imperial-oriented interests in the metropole'.[32] Hence, focusing on the consulting engineers in Westminster provides an effective historiographical path for exploring the ways in which influences were negotiated at the metropolitan end of the imperial relationship and from there redirected onto the world that the British engineering profession constantly engaged with. Moreover, the consulting engineers in Westminster constituted a group in the engineering profession that developed particularly strong imperial identities and which exerted a strong influence on other segments of the British engineering profession. These are important reasons why they are the main subjects of this book.

Consulting engineers, however, also merit attention for other reasons. The late Victorian era was 'the age of the consulting engineer', as L. T. C. Rolt rightly asserted more than four decades ago.[33] Indeed, the consultants were the leading figures in the civil engineering profession at the time. Yet, compared with the early and mid-Victorian civil engineers, scholars have devoted much less attention to this later generation of leading figures in the civil engineering profession. More recently, however, there has been a growing willingness to move beyond and do away with the alleged watershed of 1860 marked by the simultaneous deaths of the 'railway triumvirate', Joseph Locke, I. K. Brunel and Robert Stephenson.[34] Among other things this has led to fruitful reinterpretations and revisions of the notion of 'Britain's industrial decline' and an insistence that 'men of technology' have occupied complex positions in British culture that are not captivated by discourses of decline and ascent.[35] This study of empire and the engineers engages with and argues along the lines of these revisions. More specifically the book addresses issues relating to the role and status of engineers within British elite and popular culture,[36] to the 'gentrification' of engineers,[37] and to the reading cultures and communication circles that developed around the profession.[38] In relation to these issues the academic literature is especially patchy with regard to the closing decades of the long nineteenth century and the analyses therefore fill critical gaps in the scholarly literature.

Africa and British Engineers

Africa during the decades of the scramble has been subject of substantial interest in particular from scholars concerned with the causes of European expansion during 'new imperialism' and with the lasting legacies of Europe's colonial empires. In the context of this book the case of Africa is important because it allows for the exploration of how an established engineering centre in Westminster developed connections with new areas of the world and how existing systems were geared to encompass projects in these regions. In Africa the regions were characterized by great diversities with respect to environmental, economic, cultural, social and political factors.[39] Of particular importance for the concerns of this book is the different ways in which the regions were or became part of the political structure of the British empire during this period. This decided with what other groups the engineers developed connections. In relation to projects dealt with in this book the client who hired engineers was the British government either through the Foreign Office, in the case of the Uganda Railway, or through the Colonial Office via a quasi-governmental system of Crown Agents in the case of railways and harbours in the West African Crown Colonies. For the railways in Rhodesia the engineers depended on connections with the British South Africa Company while for projects in the 'veiled protectorate' of Egypt and in the South African 'self-governing' colonies they were employed directly by colonial governments.[40] These differences affected profoundly the position of engineers and this fact allows this book to unravel and analyse a wide range of connections in the chapters that follow.

One critical aspect requires further introductory clarification before the analysis commences of the metropolitan connections between British imperialism and engineering. Imperialism may be defined as the attempt to impose one state's dominance over other societies by assimilating them to its political, economic and cultural systems, usually with the use of force and violence. Imperialism, moreover, concerns the ideologies, languages and vocabularies that people have used to organize and make sense of their political, social and cultural worlds – vocabularies that have served to establish and sustain hierarchies across cultures.[41] The engineering projects that are discussed in this book were carried out in Africa at the time when the continent was partitioned by European imperial powers and during a period in which hierarchical, racialist vocabularies and visions of human society proliferated and hardened up. The works that engineers were involved with were bound to be imperial also and this raises the question from what perspective we are to view the role that engineers played in the process. According to Angus Buchanan the diasporas of British engineers in Africa were imperial because in this region of the world 'trade tended to follow the flag, and trade generated the need for engineering works'.[42] This explanation is inadequate not only because it equates British imperialism too strongly with the formal boundaries of the British empire,[43] but also because the projects that engineers were involved with were not, as Buchanan suggests, merely aiming to facilitate existing trade after political control had been established. Headrick is more to the point when he notes, in conclusion to his extensive surveys of technology transfer in the context of European imperialism, that 'Trade did not so much follow the flag as come wrapped in it'.[44] A naval intelligence book from 1919 – tellingly the most detailed and rich contemporary source for factual information on railways and railway construction in Africa – sums up what was also involved in the engineering projects:

> The aim of a colonial railway [in Africa] is the economic development of the country it serves. This covers the subjugation of unruly tribes and the military conquest and suppression of the region in question – all these strategic and political measures being mainly preparatory, and often indispensable, steps towards the government and economic development of the new country. Cecil Rhodes once stated that 'railways in the colonies are cheaper and more efficacious than guns', and it may be taken as an axiom that railways are in most cases a far better means of settling a country than wars and military enterprises.[45]

It is not the primary concern of this book to explore how the engineering projects influenced short- and long-term developments in different areas in Africa. Yet, it is important not to lose sight of the fact that the projects organized, planned and discussed in London often involved military conquest, political subjugation and violence. This insistence is not meant to infuse a moral dimension into the study but to direct attention to the fact that the organization of projects and the discussions of engineering, imperialism and Africa took place in a historical period where civil engineering projects affected profoundly the lives of people far away from Britain; people, moreover, whose voices were not heard in the debates. It is not least because of this uneven distribution of power and speech that it mattered what engineers in London thought and did. Indeed, the ideas and opinions of Britain's imperial engineers and how they saw their role in an imperial world are important issues, not least for this reason.

Composition of the Book

Chapter 1 explores reading cultures and communication circles in the engineering profession to analyse the role of empire in particular in the engineering journals which in this period served to connect and integrate on multiple levels British engineers in Africa and the engineering profession in Britain. Chapter 2 combines perspectives from cultural and economic geography to provide a spatial analysis of the imperial engineering centre in Westminster, the district which housed the headquarters of the accredited institutions of the profession as well as substantial groups of engineers including the consulting engineers of the Great George Street Clique. The professional networks, imperial connections and colonial critics of the Great George Street Clique are the subject of Chapter 3. Chapter 4 analyses the role of empire in the ICE, the most important formal platform for the Great George Street Clique and the primary institutional body in Britain for substantial numbers of engineers in colonial diasporas. Chapter 5 examines the public dimension of imperial engineering and analyses the rise to fame of the 'explorer-engineer' and the role of consulting engineers as intermediaries between engineering projects in Africa and public spheres in Britain. Chapter 6 focuses on a controversy in the Nile Valley which sheds light on the position of engineers in elite society in Britain and which brings together a number of the agents and themes examined in the book. In the Conclusion the discussion concentrates on the nature and consequences of the imperial connections that were forged in the engineering world during this period.

1 AFRICA, IMPERIAL COMMUNICATION AND THE ENGINEERING PRESS

The engineer is no mere technician. In the new gospel of industrial awakening the engineer is the missionary. The mechanic, not the farmer, is the modern pioneer. The hammer leads the plough into the wilderness.

Charles Buxton Going (1899)[1]

In their cultural history of technology Marsden and Smith rightly emphasize that 'one of the products of the long nineteenth century in Britain was a plethora of engineering in print'.[2] Indeed, by the late nineteenth century vibrant information milieus and communication systems had developed around the engineering profession. This chapter analyses how issues relating to the British empire were debated within these milieus and systems. The first section identifies channels through which the engineers who take centre stage in this book addressed issues of engineering and imperialism and it reconstructs a line of reasoning that enabled the members of the profession to argue for the indispensability of their contribution to British presence and future influence in Africa. The analysis then focuses on engineering periodicals – the publications that constituted the cornerstones in the vibrant communication milieu. Specifically, it analyses five influential journals and magazines edited from London, a location that in this period was emerging as the centre of an imperial press system in which news and information flowed back and forth between Britain and the colonies.[3]

Engineering periodicals are particularly important for this study because these widely circulated publications connected large communities of engineers across distances by transferring knowledge, news and opinions relevant to reading engineers. They are therefore useful for examining the dynamics of the connections that were developed between Britain and Africa in the engineering world. Moreover, questions relating to the British empire and imperialism featured prominently in the periodicals, but in very different ways. In exploring this aspect the analysis takes inspiration from studies of science in Victorian periodical literature and focuses on the *producers* as well as the *users* of the journals.[4] The producers were the publishers, editors and journalists who made a living from writing, collecting, 'repacking' and disseminating news and knowledge from the world of engineering. These agents were in a position to influence and shape continuously the outlook of engineers but have hitherto received very little attention in the historiography. The analysis also explores what functions the journals served for the users, those who bought and read the periodicals and whose interests the producers rarely could afford to ignore. By examining the dynamics between producers and users of the journals, the analysis traces the diverse ideas that circulated among engineers and the often conflicting imperial identities that the journals appealed to and further nurtured.

Civil Engineering, Commerce and Civilization

The late nineteenth century was characterized by a widespread belief in the civilizing potential of technology. With the rapid imperial expansion after 1880 into South East Asia, the South Pacific and Africa, the gap between Europe and the rest of the world seemed wider than it had in the first half of the century. Western technological superiority was – for a short period – unchallenged. This had a huge impact on European culture, provoking 'a fiercer assertion than ever before of Europe's cultural mission to be the world's engine of material progress and also its source of religious and philosophical truth.'[5] The idea of an intimate link between civilization and technology had a strong hold in Britain; 'the workshop of the World', that had carved out the biggest empire after the Industrial Revolution and the Napoleonic Wars partly thanks to a head start in the development of key 'tools of empire' such as river steamers, guns and railways.[6] Indeed, in the eyes of many late Victorians and Edwardians the level of technological capacities and the level of civilization were perceived to advance hand in hand. The superiority of technologies was frequently seen as an indication or even a proof of British cultural superiority as well. From there it was a short step to identifying the implementation of infrastructural technologies as an important component in the 'civilizing mission' of the British empire.[7]

This idea of the link between technological capacity and cultural rank influenced the outlook of many engineers with African connections in this period. Members of the profession were among the groups who strongly emphasized the civilizing potential of technology in the shape of railways, telegraphs and even guns. There was no singular and distinct vision of engineering imperialism for Africa, but the basic tenets of the gospel preached were 'civil engineering, commerce and civilization': infrastructure would create political stability, encourage investment and commerce, and thus allow the African continent with its allegedly backward populations to take its place as supplier of raw produce in a world economy based on London. From there, however, disagreements began – also among the engineers who are central to this book. When Frederic Shelford, consulting engineer to the British Government Railways in West Africa, spoke to the members of the African Society in London on 'Sierra Leone in the Making', he claimed that the great dividing line in the history of the colony was the introduction of the railway in the late 1890s. Prior to this the colony had been subjected to 'the efforts of the missionaries to impart some measure of Christianity into these pagan savages' but Shelford claimed that in spite of their efforts the colony was still 'struggling as a second-class British possession in which no one took any interest'.[8] In his analysis this only changed fundamentally when the British commenced equipping the colony with a proper infrastructure in the shape of a narrow-gauge railway – replacing the Livingstonian C of Christianity with the C of civil engineering.

Not all engineers saw things this way. Francis Fox, consulting engineer to the British South Africa Company for more than three decades, in one of his autobiographies asserted that:

> Engineers in the execution of great works visit all parts of the world, and I would ask them to uphold the truth. It is, I am sure, unnecessary to appeal to them to uphold our national honour, or to conduct themselves as gentlemen, for no English man worthy of the name would be found in wanting in these virtues; but I would ask them to go much farther than this – to use their influence to suppress drink, to protect the honour of women and the innocency [*sic*] of children, to maintain the observance of Sunday or the Lord's day, to encourage all true missionary work, and to sympathise with the missionaries themselves, who in consequence of many and great difficulties, are frequently liable to be depressed and cast down.[9]

Like many Victorian engineers Fox was a religious man. He took great pride in having served as consultant engineer to projects carried out to restore Christian buildings of worship including Winchester Cathedral, the Church of St Sophia in Constantinople and the dome of St Paul's in London. In his London offices the work day began with a communal morning prayer.[10] To Fox it was not a question of civil engineering replacing Christianity as the tool of civilization. Indeed, engineering without Christianity would be of little avail for a colonizing nation.

In fact, few engineers argued that Christianity and civil engineering were mutually exclusive but their relative importance as levers in the civilizing mission was a contested issue among the members of the profession. Guildford Molesworth, a London-based engineer with a long career in Ceylon and later chief inspector to the Uganda Railway, saw no problem and thought that civil engineering and Christianity would advance together. Instead he devoted numerous books and pamphlets to a fierce attack on what he believed was the real problem of the British empire and the engineering profession; the 'Cobdenite' influence on Britain's fiscal policies.[11] In his view the dogmas of free trade were rapidly undermining Britain's industry thereby preventing British engineers from winning contracts even within the boundaries of the empire. For Britain to excel in any kind of civilizing mission 'the insane worship of the baneful fetish of Free Trade' should be rejected, claimed Molesworth in 1902, and called for a tariff union within the empire. Only under the umbrella of such a union would Britain and the civilizing influence of her engineers thrive.[12]

Questions of economic policy constituted another contested issue among engineers with African connections. John Coode, consulting engineer to several harbours constructed in West and southern Africa in this period, did not, however, share Molesworth's pessimism. When Coode took the chair to address his peers as president of the ICE in 1889, he spoke on the subject 'British Colonies as Fields of Employment of the Civil Engineer, Past – Present – Future'. He listed the work carried out by engineers across the British empire and assured his audience:

> So long as the present dispensation may last, so long will there be a continuous progress in the science and practice of every branch of labour in the field appertaining to the Civil Engineer. Neither to the Engineer, nor indeed to any other disciple of natural science, would it seem to have been announced – I say it with all reverence – 'Thus far shalt thou go, but no farther'.[13]

In Coode's view it would take more than a failed fiscal policy to discontinue the advance made by the engineers throughout the empire. Progress was guided by deeper rules, natural laws of progress that only the Deity could alter should he wish.

Engineering Periodicals, the Empire and Africa

Leading engineers in London expressed differences of opinion in relation to the imperial role of the British engineering profession but what must also be emphasized is the diversity of the channels through which these issues were discussed. In the cases above Shelford wrote for a journal of a learned society, Molesworth authored politico-economic books and pamphlets, Fox published autobiographies and Coode spoke from the rostrum of the ICE. Civil engineers had entered the public arena where ideas and opinions about the future of the British empire were exchanged. The profession, moreover, had its own well-established communication channels with periodicals constituting the most important intellectual platforms. Indeed, technical journals, gazettes, reviews and magazines were a seminal part of the intellectual and professional life of a late-nineteenth-century engineer.[14] This periodical literature reveals much about what were the important issues for engineers and about how and among whom these issues were discussed. The most widely circulated publications counted their subscribers in thousands and every week or every month they reported and debated issues from the world of engineering – issues which during these decades very often concerned the British empire in relation to the engineering profession.

The engineering literature of the period has not been systematically explored and hitherto little has been written about the people who took part in producing it and even less about those who read it and the reasons they had for doing so. Yet, there are useful historiographical pointers about how to approach this literature. In a perceptive essay Eugene S. Ferguson has rightly pointed out that technical journals were not neutral conveyors of news from the world of engineering. Rather they were highly ideological publications that reflected the urgent agendas of the writers, editors, publishers and proprietors who produced them.[15] These agents were at the same time operating in a dynamic marketplace with many competitors also vying for the favour of potential readers and advertisers.[16] In order for their business to stay afloat and to pursue their other pressing agendas producers had to position themselves according to the perceived needs and wishes of those who invested money in the journals either as subscribers or advertisers. A constant negotiation was therefore going on between producers and users of journals and this pervaded the publications front to back. Unravelling this dynamic 'communication circle' between producers, media and users is crucial for getting at the underlying messages in the published texts.[17] Furthermore, in order to reach a deeper understanding of how the journal came across to readers, attention must be paid to all sections of the journals including advertisements and editorial announcements as a way of unravelling the many functions they served.[18]

By the last quarter of the nineteenth century numerous journals, magazines and reviews devoted to engineering, trade and industry circulated among engineers and groups affiliated with the profession. This literature was as diverse as it was vast. *Engineering Magazine*, launched in 1897 as a 'review of reviews' of technical literature, based its monthly digest of literature relevant to engineers on a list containing more than 80 different periodicals.[19] The analysis here focuses on five of the most influential of these periodicals: two competing weeklies, *Engineer* (1856–) and *Engineering* (1866–), the leading general engineering journals in the English-speaking world at the time; two competing engineering reviews, *Engineering Magazine* (1897–1923) and *Feilden's Magazine* (1899–1914),[20] both successful 'reviews of reviews' of technical literature; and finally the monthly journal *African Engineering* (1904–17),[21] which focused exclusively on African engineering affairs.

The Big Weeklies: Engineer and Engineering

Engineer was founded in 1856 in London by Edward Charles Henley, an engineer in his early thirties with a particular interest in railway investment and development.[22] He was proprietor of the successful journal for over forty years

– a position that was passed on to his son and later to his grandson. In 1858 an energetic young American, Zerah Colburn, joined the staff and was appointed editor, a chair he held until 1864. When Colburn left (to found *Engineering*) the editorship was taken over by a young Irish engineer, Vaughan Pendred. Pendred was editor for over forty years until he was succeeded in 1905 by his son Lough Pendred, who had been a full-time assistant editor since 1896 and like his father his tenure in the editor's chair also lasted over forty years. Editors produced many of the articles as well as the editorials in *Engineer*. They were also responsible for gathering contributors to the journal. Thus, they exerted much influence on the editorial line, but they were not alone in doing so. The manager, or the publisher as he was initially called, took care of the business side of the journal including the advertisement section which in this period took up two-thirds of the pages in each issue. From 1879 and for the next forty years the position of manager was filled by Sydney White, a nephew of the founder Edward Henley. *Engineer* was a London-based journal. It changed office addresses four times between 1854 and 1930 but never left the Strand between Westminster and the City.

After Zerah Colburn had resigned as editor of *Engineer* he (with financial support from Henry Bessemer) founded *Engineering* in direct competition to the product marketed by his former employer.[23] In size, format and scope the two journals resembled each other. Colburn claimed that *Engineering* was superior in all respects, but as it turned out there was room for both journals to prosper in the long term. Colburn conducted the journal until 1870 when he travelled to Belmont and committed suicide. *Engineering* was continued by Colburn's assistants William Maw and James Dredge, both engineers in their mid-twenties. They followed the course charted by Colburn and conducted the journal until they died, Dredge in 1906 and Maw in 1924. Like *Engineer*, *Engineering* was edited, published and printed in the Strand in central London.

Both journals were characterized by continuity in terms of their formats and sections. This was in part due to the fact that the key agents involved in their production held their positions for decades. Whereas the content of the journals reflected the changes that the engineering profession underwent around the turn of the twentieth century, the style and format indicate more stable reading habits in the profession. *Engineer* and *Engineering* had clearly found a formula that appealed to reading engineers.

The formats of the two journals were almost identical in this period.[24] The front page contained small advertisements from British industrial manufacturers and the first pages of each issue consisted of a section with tender invitations for orders on projects in Britain and in the colonies. This was followed by a job advertisement section where companies and governmental institutions advertised for draughtsmen, assistant engineers, lecturers for engineering colleges and business partners for projects in Britain as well as in the colonial world. A section of advertisements followed next, mainly from manufacturers offering boiler, condensing plants, locomotives and other types of heavy machinery, but also from publishers of engineering pocket books and companies specialized in packing and shipping machinery to all parts of the globe. In both journals the advertisement section took up more than 60 of the 100 pages that made up a weekly edition, thus leaving only one-third for the editorials, illustrated articles on projects and products, interviews with engineers, news from correspondents, letters to the editor and patent records that made up the rest of an issue. It is only this latter third that historians have taken into consideration when they have made use of these journals.[25] However, the sheer amount of space taken up by advertisements and the fact that they along with the tender invitations and job sections were placed on the front cover and in the first pages of the journals indicates that they were extremely important – as income generators for the journals but also important to the readers. It was in these pages that job and tender openings were advertised, here the suppliers of specific machinery could be identified, here shipping agencies capable of carrying goods overseas could be found. To many engineers this information would have been as great an attraction as reading articles about the progress of a large-scale British project in Africa where tenders already had been accepted, job openings already been filled, goods already been shipped. Reading engineers were more likely to focus on the next projects, on new opportunities arising rather than on projects where construction had commenced or terminated. The substantial tender, job and advertisement sections increase the awareness of the immediate business context wherein the British empire was felt among most British engineers. The professional life of a late-Victorian engineer was one in which the empire was present not only in abstract visions of 'civil engineering, commerce and civilization' but also as a pivotal factor in the business and career opportunities for growing sections of the profession in this period.

Engineer and *Engineering* had readers in Britain and also abroad, where both journals had sales agencies. In 1901 outside Britain *Engineering* had sales agents in several countries on the European mainland, in the United States, four agents in the Australian colonies and two in India. *Engineer* had a more widespread network of agencies. Outside Europe and the United States the journal was represented in China, Japan, Russia and had more agents in the colonies than *Engineering*; two in Canada, one in India, one in Ceylon, two in Australia, one in New Zealand, one in the Straits Settlements, two in the Cape Colony and one in Transvaal. Through the network of sales agents, engineers based abroad permanently or for longer or shorter periods could subscribe to the journal and have it delivered to outlets in the regions in which they were based. Unsurprisingly, sales agencies were established in the regions which housed the largest communities of British engineers (see Table 4.1 below).

Engineering was well represented in India and Australia while *Engineer* was delivered to more colonies via, among others, three agencies in southern Africa. News and articles about engineering in Africa also featured more prominently in *Engineer* than they did in *Engineering*. A telling example of the extensive coverage of South African affairs in *Engineer* followed in the wake of the South African War (1899–1902) when the journal employed a special reporter to travel through the region and comment on business opportunities opening for British engineers after the fighting had terminated. The articles that the reporter produced were published as a series under the headline 'South Africa from an Engineer's Point of View'.[26] When *Engineer* began this series in the summer of 1902, an editorial explained that the journal had made an arrangement 'with an engineer of wide experience' to make it his business to travel through all British possessions south of the Zambezi 'for the purpose of collecting information bearing upon the trades, professions, and interests of which this journal is representative'. The editorial emphasized that in 'the re-opening of a great field for trade such as South Africa, it is above all things necessary that the engineers, manufacturers and traders of the country should be placed in possession from time to time of information absolutely trustworthy' and underlined that this mission was also in line with the patriotic stance that the journal had always taken on South African questions:

> Deeply impressed as we have always been with the importance of securing for our manufacturers as much as possible of the trade in South Africa, we are determined to help as far as we can the attainment of this object ... the readers [should] feel assured that no pains will be spared to make our latest contribution to British trade enterprise a success.[27]

Engineer did not see itself as a neutral conveyor of news from the world of engineering and the employment of a special commissioner was cast as a patriotic gesture from a promoter of British interests.

As was often the case in *Engineer* the articles in the 'South Africa from an Engineer's Point of View' series were not signed. However, because the special reporter later transformed and expanded the articles into a book we know that it was J. Stafford Ransome, founder of the journal *African Engineering* discussed below.[28] From July 1902 and over the next ten months Ransome travelled in southern Africa in the capacity of special commissioner and contributed more than a dozen lengthy articles to *Engineer*. These dealt with agriculture, ports and railways in the South African colonies and in Rhodesia, and in particular with the mining industry on the Rand. Ransome made it plain that potential business opportunities were contingent upon the economic policy the British adopted after the war had ended: 'What steps are to be taken to secure as much as possible of the South African trade? Before a system of a Zoll Verein or imperial federation what are we to do?' he asked in the opening article to the series. However, no matter what line taken Ransome insisted that 'we cannot count our industrial chickens before they are hatched' and emphasized that in order to understand the industrial situation first-hand knowledge was needed. Ransome was moderately optimistic in his assessment of the opportunities for British engineers in South Africa but he emphasized that 'the uninformed has little hope of channelling his interests in the right direction'.[29] The engineer seeking his fortune in South Africa after the war had better read *Engineer*.

The columns of *Engineer* and *Engineering* continuously debated the African policies adopted by British governments. The journals generally advocated political initiatives that were likely to give the members of the British profession most work. This was perceived to be expansion and a forward imperial policy. In North Africa, South Africa, East Africa and West Africa, Britain needed to stand its ground and thereby pave the way for the civilizing and commerce-enhancing influences of the engineers. In an illuminative editorial arguing this point in 1884 *Engineer* complained about the lack of political commitment the British government had displayed in the aftermath of the occupation of Egypt:

> We cannot shut our eyes to the fact that Egypt is a country in which English engineers could find employment for years to come. The engineer is the great *civiliser* ... Without them nothing will be effected; and any government which shuts its eyes to this fact will commit a great mistake. With politics proper so called we have nothing to do, they fall beyond the scope and provinces of this journal; but we are very fully impressed with the conviction – forced home by the inexorable logic of facts – that the engineer is the great reformer, the great regenerator of modern times; and without entering in any way on the field of politics we must express our regret that this fact has not been more fully recognised by the government than what appears to be the case.[30]

The editorial is quoted at some length because it brings out significant points. It is exemplary of how engineers could assume the role as the most important British agents of empire in Africa. The conviction that engineers were the great civilizers and modernizers followed from an 'inexorable logic of facts' and while the journal placed this conviction beyond the world of politics, it was nevertheless a fact to take heed of when policies were formed. Where engineers thrived so did civilization and insisting on this 'fact' had, according to *Engineer*, nothing to do with politics proper. This way of depoliticizing the role of engineers in Africa was common in *Engineer* and *Engineering*. It was practically indisputable that what Africa needed was new technologies and that British engineers were the right people to provide them. If readers entertained any doubt about this they could on a weekly basis find many reassurances of the fact in the columns of both publications. The journals were in many instances critical of how the engineering profession dealt with the opportunities opening in Africa (as will be evident in later chapters) but it was canvassed as a fact that British engineers were the agents responsible for changing the African continent for the better. In the articles published in the big weeklies there was no question that the Cape–Cairo Railway was about to be an accomplished fact, that the railway in the colony of Lagos only marked the beginning of a bright commercial future in the entire Niger region,[31] that the dam at Aswan would turn a barren desert into the largest cotton field the world had ever seen,[32] and that the Uganda Railway would open East Africa to the civilizing influence of commerce.[33] The general message in the big weeklies was that if only engineers were presented with the right opportunities such developments were bound to follow.

The Sensationalist Monthlies: Engineering Magazine and Feilden's Magazine

Engineering Magazine and *Feilden's Magazine* were monthly magazines and like *Engineer* and *Engineering* they were direct competitors. Both magazines were established as 'reviews of reviews' of technical publications and offered their readers a digest of the literature on engineering, industry and trade that had been published during the latest month in Britain, the United States and on the European continent. Besides providing a condensation of this vast literature both monthlies in each issue carried articles written specifically for the magazine either by the editors or by independent authors who covered a specific area of expertise. Furthermore, the magazines commented on the issues and debates of the day in editorials. Advertisements also took up a significant part of an issue, between 15 and 20 pages of the 150 pages that a monthly edition contained.

Engineering Magazine was initially established in New York in 1890 but from 1897 a separate editorial board was set up in the Strand from where the British magazine was launched. In the first issue, *Engineering Magazine* assured the British readers that 'the editors in London work as if the office in New York does not exist' and that the new magazine therefore was entirely self-contained.[34] In reality a substantial part of the articles were written by Americans and its first editor was a young engineer from New York, Charles Buxton Going. Going was a graduate from the School of Mines at Columbia College and he had been associated with *Engineering Magazine* since 1896. From this position he was appointed managing editor of the British edition.[35] Not everybody was prepared to grant *Engineering Magazine* a status as fully British. When *Feilden's Magazine* was launched as a direct competitor in 1899 the publisher Theodore Feilden in his opening statement launched a frontal attack on the American connections of *Engineering Magazine*:

It is an astounding, not to say humiliating, fact that England, the birthplace of the great Engineering industries, the home of industrial enterprise and development, and the nation which has set the seal of progress on all others, should not have hitherto possessed a high-class representative trade-technical Engineering Magazine, having had to rely in this respect upon the industrial literature of the United States, which has successfully invaded this country, and has done so much towards popularising and fostering American specialities, to the displacement of a considerable amount of British trade. I have set out with FEILDEN'S MAGAZINE to remove the reproach.[36]

Feilden employed Charles Edgar Allen, a graduate in engineering from King's College in London and former mechanical engineer in Northampton, as managing editor.[37] *Feilden's Magazine* was published under the motto 'Militantly British'.

Beginning in January 1899 and throughout the year *Engineering Magazine* under the headline 'The Engineer and Imperialism' published what the editor of the magazine immodestly called 'the most important series of articles ever published by an engineering journal'.[38] The front page introducing the article series pictured a robust, white male dressed in Roman clothes holding a fore hammer in his left hand.[39] Two Corinthian columns encompassing the drawing further underscored the linkage to Western antiquity. The columns also established a clear connection to Simon de Passe's famous frontispiece for Francis Bacon's classic opus *Novum Organon* from 1620, the book that later generations often read as a tribute to the world-conquering potential of modern Western science and technology.[40] In the drawing in *Engineering Magazine* the vessels on the horizon were – as the sailing ships in the frontispiece for *Novum Organon* – returning to the Pillars of Hercules and the Straits of Gibraltar, the borders of the old world. A quote in the drawing read 'The tools and the Man I sing' and referred to Thomas Carlyle's famous statement in *Sartor Resartus*, 'Man is a Tool-using Animal ... Nowhere do you find him without Tools; without Tools he is nothing, with Tools he is all'.[41] By invoking Carlyle alongside a rich corn plant the image was firmly connected with Britain while symbolizing an Anglo-American alliance in the imperial vision of 'Tools and The Man'. Ballasted with the whole of the Western tradition and equipped with the technologies and tools of the Anglo-Saxon race the coal-fuelled ships were returning from the endless horizon as conquerors of the world.

The subjects dealt with in the eleven articles carried under the theme 'The Engineer and Imperialism' were written by British and American experts who dealt with topics such as 'The Development of German Shipbuilding', 'Mechanical Equipment for British Shipyards', 'Works Management in Europe and America' and 'The Fighting Engineers at Santiago'. However, according to the editor a coherent logic ran through the articles and tied them together in spite of the diversity of the subjects they treated. In the opening article to the series Going explained:

> An epoch in history creates an epoch in literature. Manilla, Omdurman, and Santiago, opening broad opportunities to civilisation and industry, open also the era of immensely fuller comprehension of the place and work of the engineer ... The matchless victories were primarily due to the engineer. They spoke the invincible power of mechanical skill as opposed to blind confidence in numbers and fanatical courage. The world is wakening, though tardily, to the fact that the engineer is the supreme factor in all material progress. 'Imperialism', 'expansion', and 'the open door' would be meaningless without his work.[42]

In Going's analysis the battles of Manila, Omdurman and Santiago were examples of how striking power in warfare gave industrialized imperial powers a decisive advantage over non-industrialized nations and empires. From this he drew a deeper lesson; it instigated a new era in history because battles now could be fought all over the globe with very little cost – as long as the leading industrial powers were not combatants. In Going's analysis the engineering superiority of the Western powers had changed the nature of international politics because it enabled them at very low cost to open all parts of the world to industry and commerce. This would be for the benefit of the industrialized countries but also for the new lands and people drawn into the industrial orbit as suppliers of raw produce and as consumers. In this emerging system the work of the engineer was of crucial importance because only 'under his development China, Soudan, the Philippines, and Spanish America will enrich their own people while affording us markets of inconceivable extent'.[43] Any resistance posed by 'barbarous and uncivilised fanatics' could be easily swept away. The real lesson to infer from the battlefield of Omdurman was paradoxically that real war, costly and with casualties on both sides of the field, was becoming a thing of the past. 'The age of military aggression is disappearing', Going claimed, and insisted that 'Industrial ascendency, not political power, is the animus of modern international struggles', which meant that

> The statesman who directs politics, and the military leader who protects the peace and maintains the order on which alone industry can be established and bear fruit, have thus become servants and minsters of the engineer, who carries on the actual wealth making work.[44]

In this analysis the imperial present and future of the industrialized countries were in the hands of the engineer who at the threshold to the twentieth century was no mere toolmaker.

Going was not alone in arguing along these lines but even in the columns of *Engineering Magazine* the lyrics of this 'new gospel of industrial awakening' could be sung in slightly different ways. In a discussion of how it related to the situation in China, John Barrett, a former United States minister to Siam, argued that Britain, America and Germany should unite in order to keep China 'intact but compliant' and the doors to the vast market 'free and unrestricted'. In pursuing this line of policy the engineer would, however, still be the key agent, Barrett explained:

> The more one travels, not only through China, but through Japan, Corea, Eastern Siberia, IndoChina, Siam, Malay Peninsula, Java, Borneo, Formosa, and the Philippines, the more he is convinced that their future progress and prosperity depend perhaps, more upon the engineer than upon the exporter, manufacturer, merchant, missionary, and diplomatist. Although my experience has been largely in diplomacy, what I have repeatedly seen has convinced me that the engineer is perhaps to-day the most essential personal element in the development of these unknown Asiatic countries. He will act and be successful where the diplomatist, missionary and businessman fail. There is eminently a practical side to his efforts which is beginning to appeal to the easygoing nature of the Asiatic.[45]

The markets and resources of China were laying idle, waiting for the hands of the civil engineer. The same was true of the scrambled-for continent of Africa: 'It is especially interesting to the engineer to note', *Engineering Magazine* commented in a discussion of a French scheme to construct a trans-Saharan railway,

> that in the attacks which are being made by nearly all the nations of Europe for permanent territorial hold in Africa, engineering work is depended upon as a more certain and enduring form of attack than military power, and that the railway, the canal and harbour are the real weapons in the conquest of a continent.[46]

Engineering Magazine left it to the hands of H. G. Prout, editor of the popular American weekly magazine *Railroad Gazette*, to analyse in further detail the role of British engineers in what he called 'The Economic Conquest of Africa'.[47] Prout, a former colonel of the Royal Engineers who based his authority on African affairs on his service with General Gordon in Egypt and the Sudan, was not particularly optimistic with regards to the 'unhealthy and thinly populated' continent as a field of commerce and industry. In Prout's view 'the natural line of attack on the continent would be to construct short railway lines from bases on the coast and into the interior' and he therefore dismissed the Cape–Cairo railway as an imperial phantasm.[48] Horizontal or vertical lines were, however, a question of method only; for Prout it remained a fact that the engineer was the key agent in the economic conquest of the African continent.

Late Victorians were split on the question of how Britain should maintain a dominant industrial position in the world. During this period a seminal question in this debate was whether Britain's industrial and political interests were best served by free trade or if protectionist policies and systems of imperial preference on exports and imports between Britain and the colonies were needed to face increased international competition, in particular from Germany and the United States. Numerous stands were adopted in this debate, particularly during and after Joseph Chamberlain's tariff reform campaign in 1903.[49] What is important for the context of this argument is that those who leaned towards arguments for free trade as well as those who favoured more protectionist lines argued for the importance of a strong and imperially minded British engineering profession.

Engineering Magazine supported the doctrines of free trade and unrestrained international competition. In an editorial review Going insisted that 'The theorem that one nation's ruin is another's gain has gone into the scrap-heap, with other dead dogmas of political economy' and he was willing (wrongly it turned out) to 'forecast even now that the time is not far off when Germany and the United States too, will acknowledge and accept the clear-eyed statesmanship of [Richard] Cobden and [John] Bright'.[50] The imperial standpoint that *Engineering Magazine* offered its readers was devoted to international free trade which was an integrated part of the journal's ideological platform. When the British edition of the magazine was launched in 1897, the editors explicitly claimed that its strength lay in that it was not 'limited by either national or geographical considerations'. It was 'a cosmopolitan magazine' and therefore published under the motto 'the world is its field'.[51] The publisher of the magazine, John R. Dunlap, in 1901 defended the international line of his magazine by arguing that

> engineering science is not circumscribed by national boundaries ... All applied science is a matter of plain business – a matter of pounds, schillings, and pence. 'The largest yield at the lowest cost' is the motto of success and the greatest engineer is he who first adopts and successfully applies the best that genius and invention have discovered – no matter whence its origin.[52]

The successful engineer had to manoeuvre in a world of international competition. If the British engineers and their American cousins dominated this world it was because they had adapted themselves best in a Darwinian game of international competition. According to Dunlap this was exactly the case:

> Obviously this [free trade] is a true Anglo-Saxon gospel – a gospel every clear-headed man is proud to own. For with British commerce unfettered and with her markets at home and abroad open wide in welcome to all comers we have the realism of that grand philosophy which has proved that the fittest will survive.[53]

The world of the imperial engineer was international and the sooner the game of competition was fought in accordance with the natural laws of free trade the better for British engineers. Because they were fittest they could stand their ground, maintain and expand their business in an imperializing world.

Not everybody shared this view. Indeed, many late Victorians and Edwardians considered that the 'dated Cobdenite formula of *laissez-faire* and non-intervention were plainly inadequate solutions' in a world of escalating global imperialism.[54] Also among engineers the internationalist, free trade point of view was challenged in discussions of the imperial present and future of the British engineering profession. The success enjoyed by the direct competitor to *Engineering Magazine*, the 'Militantly British' *Feilden's Magazine*, is strong testament to this fact. The line of *Feilden's Magazine* was protectionist, a point indicated by the front page of the first edition. It pictured Britannia, the emblem of British sovereignty, with her foot resting on a toothed wheel and her right arm placed above a world map with Britain colonies stamped out in dark red.[55] The publisher, Theodore Feilden, assured his readers in his opening statement that the magazine will

> be absolutely militantly British in tone, character, and purpose, giving voice to the most effective means of retaining and expanding our trade and upholding the supremacy of British institutions and British prestige the world over. Further, its most strenuous efforts will be directed towards cementing the ties of Empire and Union and the consolidation of trading interests between the Colonies and the Mother Country.[56]

The managing editor, Charles Edgar Allen, declared the mission of the journal to be 'the awakening of the National conscience' in the face of an international competition that grew fiercer:

> A cursory glance at the atlas of to-day will eloquently demonstrate that though we possess the biggest slice of the earth surface, we are hemmed in all round by new colonizing nations, who are not only taking up fresh spheres, but aggressively and successfully attacking us in our own colonies, and bagging the trade districts which have been all along looked upon as conservatively British.[57]

Allen emphasized that 'we know that British engineers are the first world over' but that this would be of little avail when Britain's formidable rivals were adopting protectionist policies. It was time for Britain to 'forestall the inroads into her trade and to fortify her own position at the head of the world's commerce and industry'.[58]

At the turn of the twentieth century the experience of the process of global, industrial imperialism did not only create visions of a world in which military competition was giving way to 'peaceful' industrial competition. There was much anxiety as the final pieces of the world were parcelled out in spheres of formal and informal imperial control.[59] The existence of this anxiety has not escaped the attention of historians but it was also felt in the technical literature and played into the intellectual debates among engineers at the time. Those who preferred a protectionist version of the new gospel of industrial awakening could find it in *Feilden's Magazine*. The magazine stayed in circulation until 1917 and it remained 'Militantly British' in tone and character throughout these years. Its editorials and articles advocated protectionism and a forward imperial policy for the benefit of British engineers and their business, as for example in a series of articles under the headline 'How Great Britain is Meeting Foreign Competition' that ran during 1902.[60] *Feilden's Magazine* also refused to accept foreign announcements in its commercial section, allowing only advertisements from British manufacturers.[61] The illustrations it carried were imbued with nationalistic and imperial iconography. The front cover of the March issue in 1900, for example, featured the British lion with its right forepaw firmly placed on a red-patched map of Africa.[62] The 'Militantly British' line appealed to many readers. *Feilden's Magazine* claimed to have a circulation of 10,000 and an estimated readership of 30,000.[63]

This number may have been exaggerated to attract more readers and potential advertisers. Yet, there is no question that both sensationalist engineering reviews reached large groups of manufacturers and engineers in Britain and also in the colonies. Indeed, the digest of the industrial literature they offered was according to the magazines particular useful for the engineers who were prevented from following the industrial press in Europe. In 1902 *Engineering Magazine* published a letter from Cecil R. Hillman of Sao Paulo Railway Company who had written to the magazine that:

> I think I am expressing the opinion of numerous engineers (situated as I am myself in out-of-the-way places like Brazil) when I say that this system [of a review magazine] is of the utmost service to those who are cut off from current technical literature. By means of it I can keep myself more or less up-to-date in the subjects in which I am specially interested.

Engineering Magazine claimed that 'hundreds of letters like this come to us from all parts of the world and always voluntarily'.[64] Through the reviews these engineers could follow debates in the technical press of the mother country. However, the digest they got was highly ideological as the journals took out controversial topics discussed in other publications and placed these in rhetorical and iconographic contexts that supported the conflicting visions of imperialism that the two magazines advocated.

The Specialist: African Engineering

The monthly journal *African Engineering* was put on the market in March 1905 as the first British journal devoted exclusively to engineering in Africa. Its daunting intention was to cover the continent from east to west and south to north. The first issue contained a map of Africa illustrating the distribution of correspondents affiliated with *African Engineering*. They were spread across the entire continent. In the text accompanying the map the editor explained that the *raison d'être* for the new journal was that opportunities in Africa for British engineers were growing so rapidly that it was 'impossible for anyone to assess this vast industrial chess board unless assisted by African Engineering'.[65] The founder, editor (until 1912) and driving force of the journal was Stafford Ransome, the author of the article series 'South Africa from an Engineer's Point of View' discussed above. Ransome descended from a family of Ipswich Quakers and engineers whose London-based business specialized in exporting woodworking machinery for overseas colonies. He was the second son in the family, a graduate from Rugby Public School and trained as an engineer in a French company. In 1893–8 he was in charge of A. Ransome & Co.'s overseas business and during this period he was elected a member of the ICE as well as of the Institution of Mechanical Engineers.[66] From the late 1880s he embarked on a career as journalist, pamphleteer and correspondent specializing in engineering matters. He wrote for several leading newspapers and journals including *Globe*, *Morning Post*, *Daily Express*, *Pall Mall Gazette* and *Engineer*. Ransome also authored books on *Modern Labour* (1893), *Japan in Transition* (1897) and *The Engineer in South Africa* (1903), the latter based on the articles he wrote as special correspondent for *Engineer*. It was thus a middle-aged engineer, experienced overseas manager and eloquent journalist who founded *African Engineering* in 1905. He developed a product that was successful from the outset.

African Engineering was first published as a 24-page monthly supplement to *African World* – a widely circulated weekly magazine edited in Johannesburg by the South African journalist, publicist and author Leo Weinthal.[67] Two years later *African Engineering* had broken away from *African World* to become a monthly in its own right, added eight pages to each issue, established a staffed office in Westminster and could boast of a circulation of 4,000 copies per issue.[68] It was published from London but had a more widespread network of sales agents in Africa than any of the other engineering journals. Eight agents were listed in the South African colonies, one in Bulawayo, one in Cairo, one in Lagos (via the Church Mission Society) and one in Nairobi.[69] It split from *African World* in 1906 in a happy divorce and it kept benefiting from that journal's network of reporters throughout the continent. Generally its correspondents and editorials were well informed on African affairs, in particular with regard to the mining industry on the Rand where many informants and readers must have had a particular connection either as mining engineers or suppliers for the industry.

The successful formula of *African Engineering* combined the characteristics of the other journals with a focus exclusively on Africa. Like *Engineering* and *Engineer* it contained editorials, letters from correspondents, letters to the editor, interviews with leading engineers, richly illustrated articles on spectacular British projects in Africa and detailed descriptions of specific engine designs. It had sections in which job openings in Africa were advertised and regularly ran small inserts from manufacturers in Britain looking to employ engineers as product vendors in various regions of Africa.[70] Its columns preached the familiar gospel of 'civil engineering, commerce and civilization' and insisted that engineers did not get the credit they deserved for their role in bringing civilization to Africa. In an editorial Ransome claimed that

> It is no exaggeration to say that every square inch of civilised Africa owes its existence, and in ninety-nine cases out a hundred its inception, to the engineer, and in many cases to men whose names are not, and never will be, known to the world. To the man-in-the-street 'Rhodes made South Africa', 'Cromer made Egypt' and 'Stanley, with a dash of Livingstone, discovered the whole'. In like manner, to the world, the mines are worked by a number of financial men whose names are international household words. And while we have no wish to detract from the work of the statesman and the politician who obtain all the kudos, or the financiers who make all the money, the political ability of the one and the pecuniary resources of the other would have been availing unless they had invoked the brains of the engineer to devise the practical means of progress.[71]

African Engineering, like *Engineering Magazine* and *Feilden's Magazine*, also collected and commented on news and information from a wide range of technical publications in Europe and America, but focused only on information relevant to engineers with African interests.

In addition to its specialist African platform, the journal ran an Agency and Information Department which aimed more directly to strengthen the business opportunities for British engineers in Africa while at the same time making the journal more attractive to potential subscribers and advertisers. The Agency and Information Department was established in the first year of the journal's existence and its core function was to serve as a business 'contact bureau' between, on the one hand, manufacturers and sales agents in Britain and, on the other, producers in Africa.[72] It offered to find commercial and technical representations in any part of Africa for manufacturing companies and engineers in Great Britain. Vice versa, it also offered to obtain British agencies for firms and individuals located in Africa.[73] In December 1906 the columns of *African Engineering* claimed that the Agency and Information Department was able to procure contact with more than 200 British firms in Africa.[74] The agency was run from the office in Westminster by two consulting engineers 'suitable for the capacity of directing enquires for the service'.[75] In serving this function *African Engineering* benefited from its widespread network of correspondents and associates in Africa as well as from the location of its headquarters in central Westminster where important segments of the engineering profession were congregated (see Chapter 2).[76]

In an advertisement pamphlet issued to attract new subscribers to the journal, Ransome claimed that *African Engineering* was indispensable to anyone who wished 'to follow and understand the industrial evolution of the continent' and more specifically

> to the consulting engineer because it affords a current record of the most recent engineering work throughout the continent ... to the manufacturing engineer because it shows him exactly what is being made for Africa ... to the commercial engineer because it tells him where to look for orders and how to get them.[77]

For engineers interested mainly in African affairs a journal devoted exclusively to this continent proved just as attractive as the larger and more expensive journals.

In 1907 *African Engineering* was equipped with a new logo and motto that aptly captivated its ideological platform and the ideas of imperial connections that the journal subscribed to.[78] Ransome, who also published as a cartoonist, in all likelihood designed it.[79] The logo pictured the British Isles in bright colours and oversize compared with mainland Europe, depicted as a small dark continent. From London rays beamed down to the African landmass which was also kept in bright colours. From the bright African continent rays also beamed back to the British Isles and written across it was the title and motto of the journal, *African Engineering: The Advocate of the British Manufacturer*. In a short text explaining the symbolism of the logo, the editor declared that the idea was not to suggest that the journal had brought light to Africa. Rather, Britain and Africa were both brightly coloured in order to highlight the connection between the two regions, because it was exactly those 'connections that it had always been and would remain the objective of African Engineering to strengthen'.[80] The producers of *African Engineering* claimed to reinforce these ties through the Agency and Information Department and in the columns that every month transferred news and information back and forth between Britain and British engineers in Africa.

'The Advocate of British Manufacturers' honoured its motto. In the first issue Ransome assured his readers that 'every word that finds its way into these columns will be written by men who have the true interest of the British manufacturer at heart, and [it] will be based on the British standpoint'.[81] As an advocate of the British manufacturer it sought to promote the interest of British engineers and constantly belittled non-British and in particular German machinery (and colonial policies) in Africa, even inviting readers to supply information on defective German machinery used in Africa.[82] Like *Feilden's Magazine*, *African Engineering* only accepted British advertisements and though its tone was less militantly British it also argued for protectionist policies to secure British stakes in the economic conquest of Africa.

However, in spite of the declared 'British standpoint', the columns of *African Engineering* often aired criticisms of London's influence on African engineering affairs. In particular, it criticized the fact that government contracts for projects in Africa were secured by an old-boy network consisting of consulting engineers in London and manufacturers in south-east England.[83] According to *African Engineering* government contracts were awarded at the expense of highly competitive British manufacturers and engineers based in Africa. The exclusion of African-based British engineers from government contracts was also a subject of complaint in the journal's 'Letters to the Editor' section, where British manufacturers and engineers stationed in Africa grumbled that contract specifications for government projects plainly favoured domestic producers.[84] *African Engineering* supported the critics. In the eyes of Ransome the best way to secure Britain's industrial hold of Africa in the long term was to support, develop and expand networks of manufacturers and engineers throughout the continent. It was to the British engineer based in Africa, Ransome argued, 'who knows what he wants' and 'who bears the consequences', that the future belonged.[85] Excluding these engineers from government contracts was counterproductive to British prospects in Africa, therefore. In the imperial vision of *African Engineering* the community of engineers in the metropolis would gradually hand over responsibility to their migrating British peers as they gained footholds across Africa.

In July 1910 Ransome launched an associated journal titled *Eastern Engineering*, along the same lines as *African Engineering*, covering the Middle East and Far East.[86] Two years later he resigned from the journal to become secretary and daily leader of the British Engineers Association, an unsuccessful lobbyist organization established with the aim of improving British export performances with China.[87] The editor's chair of *African Engineering* was left to Ransome's assistant editor Hubert Dunell, a mechanical engineer and journalist.[88] Dunell conducted the journals until 1914 but at the commencement of the First World War *Eastern Engineering* and *African Engineering* amalgamated into one journal titled *African and Eastern Engineering.*[89] It was dissolved in 1917.

Conclusions

In his global history of empires John Darwin notes:

> To an extent inconceivable as late as 1860, the world of 1900 was an imperial world: of territorial empires spreading across much of the globe; of informal empires of trade, unequal treaties and extraterritorial privilege (for Europeans) – and garrisons and gunboats to enforce it – over the rest.[90]

The scope and rapidity of this change occurring within just one generation had a huge impact on the British engineering profession. Engineers were in the midst of an escalating global imperialism and they reflected upon, discussed and wrote about what this entailed for their profession. Unsurprisingly, this development did not produce a singular or coherent engineering vision of empire but gave rise to a range of views, responses and exchanges. The analysis of the leading engineering journals makes evident how these exchanges were further shaped by the media in which they took place.

News from colonial regions, tender invitations, job announcements and advertisements from manufacturers and other agents made these publications attractive to the members of the profession. The often conflicting ideas and opinions about the political and economic course of the British empire were also part of the different packages that the journals offered their readerships. The periodicals were a seminal source of information for British engineers in the home country as well as overseas. In serving this function the journals were instrumental in tying groups of engineers together by establishing a shared frame of reference particularly among engineers in Britain and in the colonies. The periodicals were by no means exclusively imperial. In particular the big weeklies, *Engineer* and *Engineering*, covered extensively projects and issues relating to engineering in regions outside Britain and the empire where the journals also had correspondents, sales agents and many readers. However, also in these journals the empire featured prominently in the articles, news, job and tender columns, as well as in the substantial advertisement sections and the sense of community, and the identities they nurtured were in most cases imagined along imperial lines. Indeed, the success the journals enjoyed is strong testament to the fact that they served the needs of a profession to which the empire was of great importance.

The producers of journals 'wielded little authority but possessed great power'[91] as they used their publications to formulate opinions and mould thought. In particular in the sensationalist monthlies, editors and managers were very visible in the publications both in terms of photographs and with respect to signing articles and editorials. Hence it is possible to know more about them. In this period those who produced the journals were in most cases trained engineers who turned to journalism, editing and publishing. For some engineers this was an attractive career opportunity. Stafford Ransome, for example, was well educated, had worked as a manager abroad and travelled widely. He was, however, the second son in the family of engineers and his elder brother entered the family business to become manager, possibly making it more attractive for the talented younger brother to pursue a career as journalist and publisher.[92] Historians have hitherto hardly paid any attention to these influential journalist-engineers who counted their readers in thousands on a weekly or monthly basis. The journals they conducted commented on the issues of the day and had the potential to exert continuous influence on the professionals who read them. Moreover, the producers of the journals usually had urgent financial and ideological agendas and in this period often entertained strong opinions about the imperial role of the engineering profession.

The visions of engineering imperialism for Africa that the journals formulated and dispersed were diverse but they all claimed to promote British interests.
Indeed, what it meant to occupy the 'British standpoint' was a contested issue.
One avenue was to favour free trade and open international competition.
According to *Engineering Magazine* this course would be the best for Britain
and British engineers because they were fittest in an industrial, imperial game
played by the Darwinian laws of free trade. In *Feilden's Magazine* this vision was
dismissed not only as a mere phantasm but as plainly harmful to the position of
British engineers in the world. Protectionism and imperial trade unions were
needed at least as long as Britain's 'formidable rivals' did not play by the rules of
free trade. *African Engineering* was in agreement with *Feilden's* on this point but
insisted that the ties that bound engineers in Africa and Britain together should
give more support and leeway to engineers and manufactures on the ground in
Africa, when necessary at the expense of London-based engineers. There were,
however, underlying similarities in the lines that the journals adopted which
also were evident in the opinions expressed by Shelford, Molesworth, Fox and
Coode, the engineers with whom this chapter began. They all adhered to the
gospel of 'civil engineering, civilization and commerce' as the building blocks in
the conquest of Africa and to the conviction that engineers were the key agents
in a world of global, industrial imperialism. In their view engineers did not only
make tools, they also moulded the future.

2 ENGINEERS IN IMPERIAL LONDON

The Façade thus has many 'properties'; it is much more than a simple surface, covered to a greater or lesser extent in ornamentation. It characterizes a way of life. It determines urban space and its use (the 'uses' that take place in it). It has power.

Henri Lefebvre (1975)[1]

Explorations into the intersections of imperialism and urban locations have for long been a central concern in the historiography.[2] Questions of space, place and location have been tied successfully to the issue of how imperialism shaped European societies, cultures and cities. Several studies have focused on London, the largest urban centre and capital of the most extensive of the European empires. Thus, a substantial literature is available on the impact of imperialism on specific locations in London and on the activities these locations facilitated. By now a detailed 'historical map of Imperial London' is emerging from the studies.[3]

Engineers have hitherto been absent on this map and the present chapter places them there. It provides precise data on the number and location of civil engineers in London and explains why their businesses clustered around certain streets in the core of Westminster. These were also the streets where the ICE and the offices of the consulting engineers of the Great George Street Clique were located. In order to analyse and interpret this locational pattern in the engineering profession this chapter applies tools from economic geography in combination with perspectives developed in the so-called 'spatial turn' that has been influenced by post-modern and post-colonial thought. The spatial analysis of the metropolitan engineering scene brings out the growing importance that projects in the empire had to the British engineering profession. Moreover, this development not only affected the lives of the engineers who dwelled in Westminster and the activities that took place in this district. Engineers working on projects or in employment in British colonies were continuously drawn to Westminster in the course of their careers. By examining two cases of engineers engaged in what has been labelled 'imperial careering', further insights are gained into the dynamics between London and locations in Africa.[4]

The Historiographical Map of Imperial London

The scholarly interest in the mapping of imperial London is partly emerging from what has been labelled a 'spatial turn' in the social and human sciences.[5] First named and identified by Edward Soja, this spatial turn originated in post-modernist thought and received further impetus from scholars seeking to combine critical theory and cultural geography in order to overcome what they considered essentialist understandings of space and location.[6] Instead, they opted for more dynamic notions of spatiality that allowed for explorations into how place, meaning, knowledge, power relations and cultural hierarchies have been constituted through history and in contemporary society. In the re-conceptualization of space it is insisted that location, knowledge and power are inseparably interwoven. To make use of two metaphors employed by the historian of science David N. Livingstone, 'space is neither a neutral "container" in which social life is transacted' nor 'is it simply a stage on which the real action takes place'. Location is itself constitutive of systems of human interaction, it enables and constrains what can be said and done and what cannot.[7]

Europe's imperial past has been a focal point in the scholarly concern with spatiality. Post-colonial notions like Homi K. Bhaha's concept of 'in between spaces' and Edward Said's idea of 'overlapping territories and intertwined histories' have aimed to captivate how imperialism manifested itself – and still manifests itself in the post-colonial era – across space and time.[8] From this vantage point locations are studied as crossroads; hybrids with multiple identities and subject to complex processes of negotiations and interpretations.[9] This perspective has inspired a number of scholars to explore the manifestations of empire on physical locations in European cities. In the introduction to a collection of such scholarly explorations, the editors explain that the central aim is to 'explore the role of imperialism in the cultural history of the modern European Metropolis' and by insisting on the cultural hybridity of places to study 'the ways in which the experiences of empire and urbanism intersect'.[10] For the geographer Doreen Massey, the hybridity of place emphasizes that identities of locations are constituted as much by their relation with other places as by anything intrinsic to the location itself. This point Massey famously illustrated by describing a walk down Kilburn High Road in north-west London. On this street even a sketch from immediate impressions – of postboxes adorned with the letters IRA, of Indian saris, of aeroplanes approaching Heathrow from all corners of the globe – revealed ever-changing links and negotiations between Kilburn High Road and the wider world of the imperial past and post-colonial present.[11] Patterns of global interconnections were also part of the identity of the streets, offices and institutions in Westminster where Britain's imperial engineers were congregated in the last quarter of the nineteenth century.

The historical map of imperial London has become still more detailed. In an important contribution, Jonathan Schneer analysed the intersection of urban life in London and British imperialism in the specific year 1900. He examined London's docklands, using the term 'The Nexus of Empire' to capture the diverse activities and interpretations that were played out in the docks where workers, union leaders and employers handled commodities and people coming in from all corners of the empire. Schneer's general claim is 'that imperialism was central to the city's character in 1900, apparent in its workplaces, its venues of entertainment, its physical geography, its very skyline; apparent, too, in the attitudes of Londoners themselves'.[12]

The City, the financial centre of London and of the British empire, is another central location which for long has attracted the attention of historians.[13] It was in the City – situated west of the docklands and at the time simply referred to in telling geographical terms as the 'Square Mile' – that 'gentlemanly capitalists' and many others who were neither gentlemen nor capitalists met and conducted an array of financial transactions whose effects were felt across the globe. The key financial agents and institutions in the City were closely linked to the policymakers in the corridors and offices in Westminster a mile further up the Thames; closely linked by political priorities and 'gentlemanly' values but also in geographical terms. Indeed, according to Cain and Hopkins, 'the bonds between the world of finance and of politics and power were strengthened by daily contact in the metropolis'.[14] Social, cultural and geographical closeness were intertwined.

As an imperial location, the political and governmental centre in Westminster has also attracted the attention of historians. In his architectural history of *Imperial London* Michael H. Port has argued that in the second half of the nineteenth century there was a general sense that 'thinking imperially was necessary if London was to continue to hold up its head as a city worthy of its significance in the world'.[15] Port analysed how editors, opinion makers, politicians, architects and city planners 'thought imperially' when they promoted, planned, designed and constructed the streets and buildings of Westminster where the central institutions of the growing imperial and colonial administration were located. The design and construction preferences during the era of 'New Imperialism' displayed a clear agenda to rebuild Westminster to make it fit London's role as the centre of the world's grandest empire. Visions of empire – balanced against fears of ratepayers' displeasure – were shaping a massive restructuring of Westminster over a period lasting several decades. Symbols and representations of Britain's imperial grandeur were manifested in the district's buildings and monuments. The Colonial Office – completed in Whitehall in 1868 and decorated with animal figures symbolizing the five continents over which the British flag flew – and Trafalgar Square with its statues of imperial war heroes and lions symbolizing British might formed only part of Westminster's imperial façade.[16] 'Taken as a whole', Schneer notes, 'the district simply radiated national and imperial power'.[17]

It was also in Westminster, the political heart of imperial London, that the city's engineers were congregated. Table 2.1 shows the location and number of engineers in London for four selected years.

Table 2.1: Location of Civil Engineers in London in Selected Years

Year	Total[18]	South-West London[19]	'Central Westminster'[20]
1883	600	312	232
1890	599	310	245
1901	403	238	192
1909	384	245	224

The numbers are based on *Kelly's Directories of London* which in this period under-recorded poor and working-class neighbourhoods but covered the West End and its middle-class professions with exactness, not least because customers and potential subscribers for directory services primarily dwelled in these areas.[21] The numbers for engineers are therefore reliable and accurate. Issues of locations can, however, be slippery as they rarely fit neatly within either contemporary administrative boundaries or historian's categories. If an engineering office, for example, was located just north of Trafalgar Square (as some offices were) it belonged to the EC postal district in spite of the fact that it was situated closer to 'Central Westminster' than an office in the western end of Victoria Street. However, while such borderline cases complicate things they do not influence the general validity of an unmistakable observation: there was a strong tendency for civil engineers to crowd together in Westminster and in particular in certain streets within 'Central Westminster'. In 1883, 36 per cent of all London's civil engineers were registered at addresses in this confined area. The total number of registered engineers in London fell drastically after 1900 but the number of businesses in Westminster remained almost stable so that by 1909, 66 per cent of London's registered civil engineers had their offices in Westminster.

Not all engineers with Westminster addresses were gentlemen consulting engineers running an imperial business. However, the independent consultants who constituted the top layer of the profession were also congregated in the expensive and prestigious core of Westminster. Evidence for this can be obtained from the location of the members of council of the ICE. These were the select members of the small governing body of the most high-status of the British engineering institutions. New members of council were appointed by the sitting council whose members took a keen interest in getting those considered most distinguished and worthy into the governing body. Nomination and election was a clear indication that the candidate belonged to the elite circles of the civil engineering profession.[22] As is shown in Table 2.2, the distinguished members of council were almost exclusively Westminster dwellers.

Table 2.2: Location of the Members of Council of the Institution of Civil Engineers in Selected Years[23]

Location	1885	1895	1902	1912
'Central Westminster'	26	21	22	11
London outside Westminster	3	5	3	6
Outside London	3	5	10	18
Total	32	31	35	35

Engineers in Imperial Westminster

Civil engineers were thus strongly represented in 'Central Westminster' and elite consulting engineers featured prominently within these numbers. For analysing the causes and consequences of this fact, Urban Cluster Theory is a useful historiographical tool.[24] Taking its cue from economic geography, this perspective focuses on the economic and social benefits that accrue to professions by clustering or co-locating. Urban Cluster Theory has been applied by the economic historians Michael Ball and David Sunderland in order to explain the relations between physical locations and developments in various professional groups in London in the nineteenth century.[25] It is also useful for exploring why engineers congregated in Westminster in this period. Moreover, Urban Cluster Theory constitutes a constructive supplement to the perspectives connected with the 'spatial turn' because it works at a quantitative level and therefore helps to establish general locational patterns into which specific case studies fit.

Vicinity of Imperial Politics

Generally service sector professions tend to be physically close to the functions and clients they serve. In their economic history, Ball and Sunderland demonstrated that this was an important reason why the service sector in Britain in general became concentrated in London in the nineteenth century. Among other professional groups Ball and Sunderland analysed the clustering of accountants in the City, publishers in and around Fleet Street, and the legal professions that clustered in Westminster where the central institutions of law and justice were located.[26]

The general point that clustering occurred around clients and customers also applies to civil engineers. Civil engineers first began to set up offices in Westminster in the late 1830s. The magnet that initially drew them there was the Parliament. Like most developments in the civil engineering profession in the early Victorian era this was intimately tied to the construction of the domestic railway system. Projected railway lines had to be approved by special parliamentary committees before any construction work could commence. It became the job of certain civil engineers not only to make the initial surveys for future railway lines but also to prepare plans and bills that could pass through the no-less demanding obstacles posed by the political system. For this purpose engineers set up their drawing offices within easy reach of Parliament and the Board of Trade – the latter situated in the north wing of the Treasury building and after 1875 in a new building in Whitehall Gardens next to the Colonial Office.[27]

Some engineers became experts in knowing the ins and outs of the Byzantine system of policymaking in Westminster – a system notorious for its opaqueness and for the inertia of its bureaucracy, and therefore often subjected to lampoon, as for example in Charles Dickens's *Little Dorrit* where 'the whole science of government' was contained in the phrase 'How not to do it'.[28] Manoeuvring within the political system required skill and experience, and it required *presence*. Zealous and successful younger civil engineers like John Fowler, John Hawkshaw, Francis and Douglas Fox, William Shelford and John Coode – who as will be described in the next chapter all set up independent practices in central Westminster prior to 1875 – spent a substantial part of their time in the rooms and corridors of Westminster's political institutions, lobbying and laying out plans before parliamentary hearing committees. Indeed, persuasiveness in committee rooms was a crucial factor in determining the success of a professional engineering career at the time.[29]

Contemporary voices confirmed that proximity to the politicians remained an important prerogative into the twentieth century. In 1905 Thomas Claxton Fidler, professor of engineering at Dundee University but also registered at an address in Westminster, noted that 'London is said to be the city which contains the largest number of civil engineers in proportion to its population; and their chambers are congregated mostly at Westminster within easy reach of the Houses of Parliament'.[30] Fidler went on to provide an idealized but rare firsthand description of how engineers, entrepreneurs and politicians interacted in Westminster:

> There is one season of the year when the office [of a Westminster civil engineer] shows signs of more than usual activity. Men of business and members of Parliament have come back to town after the August holidays with freshened vigour, and this is the sowing time for enterprise. Corporations, directors, and company promoters begin to think again of some cherished scheme, and they know that if anything is to be done for the next year or more, it must be put in the hand at once for the plans must be deposited in Parliament by the end of November ... There is a great deal to be done, and no time to be lost. The engineer is consulted, and a surveying party immediately takes the field to prepare the 'parliamentary survey'. After this the drawing office is busy preparing the parliamentary plans and sections. These are not technical plans; their object is to illustrate for committees and for interested parties how the project will affect the landed interest of proprietors, and all industrial, parochial, municipal, county, or imperial interests.[31]

Fidler directs attention to the many types of clients and interest groups in Westminster that it was advantageous for engineers to be in the vicinity of. In particular governmental and colonial institutions – what Fidler referred to as 'the imperial interest' – were well represented in the streets of central Westminster. Indeed, as is evident from Table 2.3, most of the imperial and colonial institutions in London were literally around the corner from the offices of civil engineers.

Table 2.3: Location of Selected Government Offices and Colonial Representations in Westminster in 1887[32]

Government Offices / Colonial Representations	Location
The Admiralty	Whitehall
Board of Trade	Whitehall Gardens
Colonial Office	Downing Street
Crown Agents for the Colonial Office	Millbank
Foreign Office	Downing Street
Home Office	Whitehall
India Office	Charles Street, Westminster
Treasury	Whitehall
War Office	Pall Mall
Office of High Commissioner of Canada	9 Victoria Chambers
Cape Colony Government Office	9 Victoria Street
Agent General for New South Wales	5 Westminster Chambers
New Zealand Government Office	7 Westminster Chambers
Queensland Government Office	1 Westminster Chambers
South Australia Government Offices	8 Victoria Chambers
Office of the Agent General for Victoria	8 Victoria Chambers

In the last quarter of the century these institutions grew in importance to the British engineering profession and in particular to the consulting engineers who were occupied increasingly with projects throughout the empire. Indeed, Parliament may have lost some of its magnetic force after the completion of the main trunk lines of the domestic railway system around 1860 but the consulting engineers stayed in the area when Westminster from this time began to change to fit its role as the administrative centre of the expanding empire. Well-informed contemporaries like Fidler noted this. So did the editors of the *Engineering Magazine* when the journal in 1904 observed that 'Great George Street is in close proximity to important government offices including: the Foreign Office, the Office of Works and the new Admiralty Building'.[33] When Africa opened as a field for imperial engineering projects, civil engineers' offices had already been situated in Westminster for a long time. The early presence of engineers in the area is particularly significant because it was engineers who had managed early on to position themselves in Westminster who secured projects in Africa after 1875 when British expansion and presence gained momentum in that region.

Vicinity of Associate Services

Governmental institutions and policymakers were not the only magnet that drew service sector professions to London. Generally professions cluster with other professions that provide ancillary or associate services.[34] This was also the case for engineers in Westminster. Running a successful career and business as a consulting engineer depended on the existence of other professional groups providing related services. Such groups were also located in and around Westminster. In this regard the contractors that engineers worked with constituted a pivotal group. Contracting companies who won tenders for large projects in Africa, as for example Lucas & Aird (37 Great George Street; Bechuanaland Railway), Ransome & Rapier (5 Westminster Chambers; Aswan Dam), Pauling & Co. Ltd (Victoria Street; Rhodesia Railways), had main or branch offices in Westminster.[35]

Contractors were big players sometimes employing thousands of people but engineers were also dependent on numerous other less visible professional groups. One example is precision instrument-makers whose congregation in the nineteenth century in Clerkenwell immediately north of Westminster has been documented by Ball and Sunderland.[36] Precision instruments were required by scientists, geographers and in particular the military (the maxim gun was invented in Clerkenwell in the early 1880s), but civil engineers formed another important customer group for instrument-makers because they needed reliable compasses, theodolites, watches and other precision tools when they designed and organized construction projects across the globe.[37] Claxton Fidler provided another telling example of a specialized and highly localized ancillary profession, that of parliamentary lithographers. Lithographer offices in Westminster had developed an expertise in transforming the materials produced in the drawing rooms of civil engineers into state-of-the-art charts and maps suited for the corridors of Whitehall and Parliament.[38] This group was endemic to the Westminster area. So were parliamentary solicitors, the juridical experts who specialized in guiding private bills through parliamentary processes.

Parliamentary lithographers and solicitors in Westminster and instrument-makers in Clerkenwell are illustrative examples of specialized and highly local ancillary services needed by engineers. Also important for the prerogative of Westminster was the fact that other branches of engineering were strongly represented in the location. There were roughly twice as many mechanical engineers registered in London as there were civil engineers and especially after 1890 the number of electrical engineers also increased rapidly.[39] Large-scale engineering projects required the services provided by specialized groups of engineers. Railways are obvious examples of collaborative works between these sub-branches. In broad terms, civil engineers planned the railway trajectories and designed the bridges, mechanical engineers were responsible for the design of suitable locomotives and machinery, and the telegraphs which were established along the tracks were the area of expertise of the growing population of electrical engineers.

Apart from the more specialized professions, civil engineers were also dependent upon the services of professionals such as bankers, solicitors and accountants in order to run their businesses. While these groups were found across the country it was rare to find so many of the ancillary professions within such a confined space as Westminster. The combined expertise of these specialized service providers were needed for the execution of engineering projects across the empire. Indeed, their mutual services were often indispensable when the politics of formal and informal imperialism were put into practice. The centrality of Westminster in imperial London owed much to the fact that these service providers clustered within a few streets in the district.

Vicinity of Information and People

The privilege of a location is often tied to the existence of efficient information channels. Relevant news might be about process innovations, changing demands in volatile markets, political intrigue or financial opportunities. Whatever the need, the imperative to cluster to obtain or share information can be great.[40]

Westminster had unrivalled formal and informal channels that facilitated the sharing not only of technical information, but also of relevant news pertaining to politics and finance. In this location information from the world of engineering and the world of imperial politics intersected. It was by no means coincidental that the journals and review magazines discussed in the preceding chapter were edited and published from offices in and around Westminster. This was the location where news could be obtained. It was also here one could meet the right people in the right places – here key agents engaged with each other and the assignments for engineering projects in the empire were negotiated. Presence and physical closeness were essential as success depended on access to the rooms and corridors where information flowed and decisions were taken. Moreover, connections between location and information were also a question of timing. Though a virtual communication revolution was taking place in this period (in the shape of ocean steamers, railways and telegraphs), information and news still travelled between London and overseas colonies with a considerable lag.[41] This aspect is of particular importance in this context because imperial politics developed rapidly and so did the opportunities for colonial engineering work in Africa. Windows of opportunities could open almost instantly and close again just as rapidly.

Westminster contained the venues for social and professional engagements between engineers and other influential groups with imperial interests. The archetypal setting for informal contact in Victorian London was the gentlemen's club.[42] Here upper-class Victorian males socialized with their peers. Many clubs were political or devoted to arts and literature while others were reserved for certain professional occupations. The most fashionable clubs were housed in stately mansions where gentlemen met in the lounges, dining rooms, smoking rooms, billiards room, cards room or the libraries and news rooms to socialize and foster new or nurse existing connections and contacts.[43] Clubs were immensely popular in late-Victorian London. *Kelly's Mayfair, St. James's, Soho and Westminster Directory* for 1887 lists 96 principal clubs and club-houses in these districts alone.[44]

It is invariably difficult to pin down in solid evidence that off-the-record club interactions and conversations mattered. Yet, in formalized and hierarchical cultures like those thriving in the West End, clubs offered a rare free space for informal engagements. And the Westminster gentlemen engineers were there, in the same rooms as politicians, financiers, scientists and men of the arts discussing the issues of the day in politics, business and culture. The most popular club among the consulting engineers of the Great George Street Clique was the 'scientific/literary' Athenaeum Club in Pall Mall. Members from the ranks of the Westminster consulting engineers during this period included William H. Barlow, Frederick Bramwell, Harrison Hayter, Benjamin Baker, John Wolfe Barry, John C. Hawkshaw, William H. Preece, Alexander M. Rendel, Alexander Kennedy and Robert Elliott-Cooper.[45] According to Ruth Barton, the Athenaeum was 'the organizational embodiment of cultural authority' among elites in Victorian London. In a study of the formation of the illustrious X-Club she has demonstrated that ambitious young scientists like T. H. Huxley, John Lubbock and Joseph Hooker in the 1850s and 1860s eagerly worked to obtain membership of this particular club. They considered the Athenaeum an important location in their pursuit of professional and gentlemanly status and saw it as a lever 'for conferring authority and respectability' onto the rising natural sciences.[46] Similar ambitions can be seen in the preference for this club displayed by the leading engineers in late-Victorian London who also yearned for a position among Britain's cultural elites – a theme to which we shall return in several occasions. Other popular clubs among engineers were politically orientated and included the two 'conservative' clubs, St. Stephen's Club in Bridge Street (members included John Fowler, Harrison Hayter, Guildford Molesworth and William Matthews) and Carlton Club in Pall Mall (members included John Fowler, Frederick Bramwell and Charles Metcalfe). The 'strictly liberal' Reform Club in Pall Mall also enjoyed some popularity among leading engineers (members included Robert Rawlinson, John Wolfe Barry, James C. Inglis and William C. Unwin). Admission to clubs was regulated by formal rules and based on personal recommendations. In this sense there was a clear spatial dimension to the question of access to information and by implication to engineering projects. Presence alone was of course not enough, as one could be in the rooms and still not be part of the conversation, but it was significant that information flowed within a very confined space to which (some) engineers had access.

While clubs were important for engineers it was the accredited institutions and associations of the profession that constituted the cornerstones in the vibrant information milieu outside the confines of the offices. Like individual engineers the headquarters of the accredited institutions congregated in and around Westminster. Indeed, professional bodies chose to settle in the neighbourhoods of Westminster throughout the period covered in this book.[47] The gathering and dissemination of knowledge from the world of engineering through lectures, conferences and meetings were among the core functions of the accredited institutions. Chapter 4 examines in detail the role of the ICE in this regard but the agenda to accumulate and disseminate information was shared by practically all professional engineering bodies in Westminster. The Institution of Electrical Engineers, located on the Thames Embankment, in a typical manner declared its objective to be 'the promotion of electrical and telegraphic science and its application, and to facilitate the exchange of information and ideas on these subjects by means of meetings, exhibitions, publications, and the establishment of libraries'.[48]

The facilitation of informal contact and communication between peers was among the core functions of the accredited bodies and this was a main reason why they clustered in Westminster. To honour declared aims such as that 'to enable Mechanical Engineers to meet and to correspond, and to facilitate the interchange of ideas' (Institution of Mechanical Engineers),[49] 'to afford facilities for education, study, and self-culture to Marine Engineers' (Institute of Marine Engineers),[50] or to encourage 'social and friendly intercourse among members and visitors' (Society of Engineers),[51] presence in central London was imperative.

Not all information was confined to Westminster. As described in the preceding chapter, printed material in the shape of technical journals and engineering reviews were distributed across the Britain and the empire via the postal systems and networks of sales agents. The largest engineering institutions also printed for distribution the papers that members presented at meetings in the lecture theatres in Westminster. Yet, news and information exchanged, for example, in informal club conversations, in the council room of the ICE and in engineers' offices were tied to specific locations. This information was in many instances the most important; because it was local it was also unique, available only to insiders, fixed in the circles of people who walked the streets, frequented the halls, corridors and rooms in Westminster.

Headquarters

For accredited associations a central address in the capital was a way to increase public recognition and respectability.[52] Presence in the metropolitan centre displayed power. A strong indication of the high priority attributed to Westminster by the engineering institutions – and by implication by the members they served – can be identified by examining the changing locality of the headquarters of the accredited bodies of the largest groups of engineers, respectively electrical engineers, mechanical engineers and most importantly civil engineers.

The Institution of Electrical Engineers was founded in 1871 (until 1888 under the name the Society of Telegraph Engineers). Its first headquarters were established in 2 Westminster Chambers. In 1874 the electrical engineers relocated to Broad Sanctuary in the south end of Whitehall. In 1891 the expanding institution (its membership by then numbered nearly 2,000) transferred its offices and library to Victoria Street. A permanent solution to the housing problem was, however, not arrived at until 1904 when the electrical engineers purchased a lease from the Royal College of Physicians for Savoy Place – a massive building situated on the Victoria Embankment just north of Whitehall.[53]

The Institution of Mechanical Engineers displayed a similar pattern of relocation within certain streets in Westminster. The mechanical engineers had established their professional institution in Birmingham in 1847 partly in opposition to the London-based ICE.[54] In 1877, however, the members voted that the headquarters be relocated to Westminster. They were initially established in 10 Victoria Chambers but the search for more suitable accommodation elsewhere in Westminster intensified as the membership grew along with the economic strength of the institution. A lasting solution was found in 1895 when a property in Storey Gate just off Great George Street was secured. Like the headquarters of the electrical engineers this building was stately, impressive and located in the heart of Westminster.[55] As R. H. Parsons noted in 1947 in his centenary history of 'the mechanics':

> The position was excellent from every point of view. In the heart of that part of London immemorially associated with the engineering profession, but it had the further attraction of overlooking St. James's Park while its dignity was enhanced by the proximity of many of the most important government offices.[56]

The attraction of and attachment to Westminster were even more pronounced with regard to the ICE, the oldest of the professional engineering associations in Britain. Founded in 1818, the institution moved into a building in 25 Great George Street in 1839. Initially these were modest headquarters but they gradually underwent enlargement. In 1846 a new lecture theatre was added and in 1866 the ICE expanded its headquarters by securing the rear of 24 Great George

Street. In the early 1880s, 26 Great George Street was also purchased and the ICE commenced with the erection of a new building on the premises of 24–6 Great George Street. These headquarters were completed for the opening sessions of 1895–6.[57]

The establishment of the new building for the ICE occurred at a time when British governments designed and constructed the new offices in and around Whitehall to accommodate the growing national and imperial administration. This rebuilding lasted over four decades and amounted to an almost complete restructuring of central Westminster.[58] Great George Street was the main thoroughfare between Parliament Street and St James's Park and therefore indispensable to any scheme to build between the Thames and the parks. The home of the ICE barred the westward expansion of the government offices and it thus occupied an exposed but also potentially lucrative spot. The council of the ICE was unwilling to relinquish its advantageous location. From the 1870s the council regularly fended off government attempts for compulsory purchase of the premises. In 1892, the government issued new plans for 'Improvement of Parliamentary Street' and as these would result in the relocation of the ICE the council petitioned against the scheme.[59] The petition stated that 'The Institution is the head and centre of the science of civil engineering in the United Kingdom, India and the Colonies' and it insisted that 'the site your petitioners occupy is especially suited to their requirements' and that 'any interference would impede their abilities to pursue objects for which they have been given a charter'.[60] By 1900 the ICE occupied the only significant property on the north side of Great George Street still in private ownership.

After 1900 confidential and protracted negotiations between the council and the government led to an agreement that was highly favourable for the ICE. The council assented to give up its building from 1895 and in return the civils were accommodated with a property in 1 Great George Street on the south side of the street. In addition, the engineers were awarded a financial compensation that abetted the ICE in constructing a new and larger building that was completed in 1912.[61] The difference between the three buildings occupied by the civils over this period is striking and the change bears testament to the rising economic power and social standing of the profession that took place during these decades. The ICE went from occupying a standard three-storey Victorian house to residing in a custom-built 25,000-square-foot building fully equipped with ionic marble columns, stately chandeliers, mahogany staircases, luxurious committee rooms and spacious lecture halls.

In an in-depth study of the new ICE building and its 'paraphernalia' Malcolm Dunkeld has shown that 'the architecture of the building mainly reflected the social, cultural and aesthetic assumptions' of the London consulting engineers who dominated the council of the ICE as well as the building committee that was established for planning and overseeing all the phases of construction.[62] And it was an imperial building that these engineers chose to erect at this point in history. At the ceremony for laying the foundation stone for the new head-quarters in 1910, the president of the ICE, Westminster consulting engineer James Charles Inglis, declared:

> We feel that we have an Imperial function to fulfil. There are more than 2,500 of our members who reside in distant lands, and we feel that as year after year passes they are contributing a greater proportion of knowledge of the facts and of experience to our proceedings (Hear, hear). I should say that half the papers that are published are descriptions of works beyond the sea. Therefore we must look after them. We on this island hope that the Institution will always be the parent Engineering Institution for British Engineers all over the Empire ... It is in that spirit that this stone has been laid, and whatever the difficulties of the moment, we as Civil Engineers, taking a general view of our civilisation and our Empire, feel that in laying this stone we are laying a stone of Empire, laying something to provide a place of work not only for the present generation of engineers but also for those who will come after them.[63]

The new home for the ICE was from the very foundation inscribed with an imperial identity. This was also commented upon in the press. In an article on the new building *African Engineering* noted that

> An interesting feature in the construction of the building will be the introduction of the timbers and certain other products of various parts of the British Empire, which may serve to symbolise in some degree the close relationship of those who constitute the Institution of Civil Engineers in all parts of the world.[64]

In his study of the ICE building Dunkeld has not found evidence that the building committee enforced rigorously such an imperial purchasing policy with respect to construction materials. However, his findings show that the orna-mentations of the building were imbued with representations of the British empire and that keen attention to detail was paid in selecting these adornments. The seven decorative panels facing Great George Street were ornamented with floral emblems representative of Britain, the Dominions, the Crown Colonies and India; the Tudor rose, thistle and daffodil representing Britain (including Ireland); the maple leaf for Canada; for New Zealand the silver fern; a protea species for South Africa; the golden wattle for Australia; a palm leaf for all the Crown Colonies and lotus flowers to represent India. When selecting these emblems the secretary of the ICE communicated extensively with high commis-sioners for the Dominions, the India Office and the Colonial Office for advice on appropriate flower emblems for the different territories. It proved particularly difficult to settle on a flower that could represent India and the ICE received several suggestions from the individuals consulted. The lotus was finally chosen after it had been recommended by Sir Christopher M. G. Birdwood, a former colonial administrator and special assistant in the statistical department of the India Office, who suggested that the lotus could be said to represent Hindu as well as Muslim communities.[65]

Another diplomatic concern in relation to the façade ornamentation was how close the different emblems were to be to the United Kingdom insignia that was to be placed above the main entrance. As Dunkeld explains the concern was:

> That the position of the emblems could offend ICE Members, the government and/ or the respective Dominions; the issue at stake was the degree of symbiotic closeness or emotional bond that existed between the motherland and the Commonwealth and how this should be represented in actual distance on the exterior of the building.[66]

Faced with this intricate problem the building committee asked the Colonial Office to determine what could be considered a 'neutral' order of precedence. This was clearly a pre-emptive manoeuvre in the event that colonial ICE members or business partners and political associates in Westminster would express dissatisfaction with the finer details of imperial precedence expressed on the frontage of the new headquarters.

The new ICE building thus expressed a specific, symbolic materialization of the empire. It involved the construction of a building but it was also a tangible, physical manifestation of the idea that the engineering profession was an imperial profession. The new home of civil engineers aimed to encompass an imperial present but it was also pointing forward to an imperial future for the profession. These were to be the halls and rooms to accommodate the next generations of Britain's imperial engineers. Like the Colonial Office, the Foreign Office and the India Office further down the street the new home of the civil engineers was an integral part of imperial Westminster. Civil engineers not only dwelled in the centre of imperial London, they actively contributed to the symbolic and physical construction of it.

Westminster and Colonial Engineers

As noted above the perspectives developed in the 'spatial turn' and in particular the insistence on the 'hybridity of place' have drawn historians' attention to the fact that identities of locations are constituted as much by their relation with other places as by anything intrinsic to the location itself. This was also the case for the engineering centre of imperial Westminster. A path for exploring this is to focus on how engineers engaged in projects across the empire were connected with London. Indeed, civil engineers pursuing careers in the colonial world were continuously drawn into Westminster during the course of their lives. By focusing on two exemplary cases this section identifies and analyses dynamics between the metropolitan engineering centre in Westminster and the areas in which colonial projects were set.

George Whitehouse was chief engineer to the Uganda Railway constructed between Mombasa and Lake Victoria in the period 1896–1903.[67] Born in Cornwall in 1857 Whitehouse received his civil engineering degree in 1877 from King's College, situated just north of central Westminster. Shortly thereafter he obtained his first of many employments as railway engineer on a British project overseas. His first job was with the Durban–Maritzburg Railway in Natal. His next assignment was with a project in Mexico where he worked on a mountain section of the national railway between Vera Cruz and Mexico City. After a short spell in Britain on the Chatham and Dover line he went overseas again, this time to India, where he was employed by the public works department in the North-Western Province also on the establishment of a mountain railway. From there his next stop was South America, this time engaged in the construction of a railway in the Peruvian Andes. He became member of the ICE in 1892.[68] Whitehouse was appointed chief engineer to the Uganda Railway in September 1895 and held the position until May 1903.[69]

During his years as chief engineer to the Uganda Railway Whitehouse was based in East Africa with his family and made two journeys back to England. His diaries show that he was in London from August to November 1897 and again from April to September 1900.[70] When Whitehouse was back in Britain he spent his days visiting relatives while recuperating from the malarial fever that he had contracted in East Africa. He was also working during his brief visits to his mother country. When doing so he was in Westminster. His diary shows where he went and why and to exemplify this we may focus in on his whereabouts during three consecutive days in June 1900.[71] On 18 June Whitehouse attended a meeting with the Uganda Railway Committee at the Foreign Office. This select committee had been set up by the Foreign Office in 1895 when the British government had decided to construct a railway from the Indian Ocean to Lake Victoria.[72] The committee consisted of Clement Hill (head of the Foreign Office's Africa Department), Francis Bertie (Foreign Office), Sir Montagu Ommaney (of the Crown Agents for the Colonies), G. L. Ryder (the Treasury), Sir John Kirk (former governor-general of Zanzibar), Alexander M. Rendel (consulting engineer to the Indian State Railways) and Francis O'Callaghan (civil engineer and former secretary of the Public Works Department of India). Francis O'Callaghan was managing director and in charge of the Uganda Railway Committee's daily operations.[73] He was also George Whitehouse's uncle.[74] O'Callaghan made his way to a high position in the Foreign Office through his lifelong affiliation with the establishment of the railway system in India including the mountain railway in India that George Whitehouse had worked on at the onset of his colonial career.[75]

The committee sitting in Westminster was in charge of the Uganda Railway project during the years of construction. As chief engineer Whitehouse was head of operations in East Africa but he was obligated to clear any major decisions concerning the railway with his superiors in London. When he was in Africa he stayed in contact with the committee via telegraph and the mail boat service but communication was much more efficient during his stays in Britain. Major decisions were therefore addressed at these times. At the meeting on 18 June it was discussed how to deal with a parliamentary hearing on the Uganda Railway that was scheduled to take place two days later in the House of Commons, situated further down the street from the Foreign Office. The committee anticipated (correctly it turned out) severe criticism in the Commons as the railway project had overstretched its budget and needed Parliament to sanction a further cash injection of at least £2 million directly from the Exchequer.[76]

The next day, 19 June 1900, Whitehouse spent his day just a stone's throw from the Foreign Office, in the office of Sir Alexander M. Rendel located in 3 Great George Street. Rendel had been consulting engineer to the Indian State Railways for more than three decades and was also employed as London consultant to the project in Uganda. In his diary Whitehouse recorded that the issues addressed that day were the designs of the steel viaducts needed for scaling the Kikuyu Escarpment and the Mau Escarpment forming the eastern and western rim of the Rift Valley that the railway traversed on its way to Lake Victoria. With a height of nearly 10,000 feet the steep cliffs of the Kikuyu and Mau Escarpments posed the biggest topographical challenges to engineers of the Uganda Railway. It was on account of this natural obstacle that it was decided in the summer of 1899 to place the new East African railway headquarters in a swampy area then known as Nyrobi Hill and today the site of the city of Nairobi.[77] In the spring of 1900 Whitehouse and a group of engineers had carried out a survey of the escarpments[78] and back in London he in collaboration with Rendel's office drew up the specifications for the dozens of steel viaducts needed for getting the tracks down the valley and up again. In the evening Whitehouse dined at the Carlton Club with his uncle and boss, Francis O'Callaghan.

On 20 June Whitehouse was in the Palace of Westminster attending the parliamentary hearing on the Uganda Railway in the House of Commons. O'Callaghan gave evidence. The debate was heated as the £3 million originally allocated for the railway had been spent and the railhead was only roughly midway between the Indian Ocean and Lake Victoria. In 1900 the Conservatives held the majority in Parliament and in spite of the criticism directed at the railway project in general and the committee in particular the bill allocating extra funds for the railway was eventually passed with a considerable margin. By the time the railway was completed in 1903 the total expenditure was at least £5.5 million.[79]

Whitehouse's whereabouts during these three days in June were typical for his stays in London. During the summer and autumn of 1900 he regularly attended meetings in the Foreign Office and worked in Rendel's consultancy offices. The issues that required further attention included the viaducts for the Rift Valley, water supplies for the newly established railway headquarters at Nyrobi Hill and the settlement of a strike that had broken out among the Indian railway workers in East Africa. On 6 September Whitehouse returned to Africa.[80]

Whitehouse's movements in central Westminster are illustrative of the fact that the key metropolitan institutions where policymakers, civil servants and engineers negotiated the political and technical solutions to imperial engineering projects were located within a very small area. Government offices, the Houses of Parliament and the offices of engineers were literary situated next door from each other. It was in these locations that the political agendas materialized into projects carried out on the ground in Africa. Indeed, within this confined space decisions were taken that had far-reaching consequences throughout the globe – in this particular case in East Africa. It was, for example, within the circle of engineers and administrators who had made their career constructing railways in India that the decision was taken to import labourers from Punjab for the construction of Uganda Railway. By the time the railway was completed more than 30,000 Indian workers had been recruited and transported to East Africa.[81] Many of those who survived the construction period stayed on after the project was completed and made up an influential segment in the Indian expatriate community that was to play an important role in the history of East Africa in the twentieth century.[82]

Decisions such as that to import Indian labour or to place the railway headquarters at Nyrobi Hill were taken in the offices and corridors of Westminster. Though developments were not determined or controlled from Westminster, the choices made in that location brought with them large and often unforeseen consequences. Such instances show that in this period London's influence was felt inland in East Africa in ways that it had not been previously. These are tangible examples of what Sandip Hazareesingh has labeled 'interconnected synchronicity' to denote instances 'whereby "imperial" causation occurs in both arenas, through a set of related, continuous, and mutually transformative processes'.[83] Indeed, they bear testament not only to the growing interconnectedness between Westminster and East Africa but also to the uneven distribution of power that is characteristic and constitutive of connectedness within an imperial frame.

George Whitehouse was a middle-aged, upper-class white male. He was engineer-in-chief to a large and prestigious imperial project in Africa. Upon the completion of the Uganda Railway he was created Knight Commander of the Order of the Bath (KCB).[84] He belonged to a privileged group of British engineers with access to the halls and offices in Westminster. Yet, for Whitehouse

London was as much a place of subjection as it was of power. He was in charge of the organization in East Africa for seven years but in Westminster he was in some respects a marginalized figure. He had only been the Uganda Railway Committee's third choice for the job of chief engineer and accepted the position at an annual salary of £1,200. This was £200 less than had been offered to the other candidates.[85] Throughout the years of construction Whitehouse's relationship with the committee in London was tense and it kept changing for the worse. There were several reasons for this, one of which was financial. Although his salary had gone up to £1,500 by 1898, his diary reveals that he felt his wages compensated him inadequately for his trials and responsibilities in East Africa.[86] Moreover, Whitehouse was at times scolded for not following procedures staked out from London. He was, among other things, reprimanded for purchasing coal in India without the consent of the committee and on another occasion he was forced to withdraw a proposal he had put forward for improving the leave terms for railway staff in senior and junior positions – a proposal the committee considered too generous and expensive.[87] In addition, short on money and time, the committee was impatient with the progress of the railway. After 1901, in particular, Whitehouse was put under pressure from London to get the railway more speedily to Lake Victoria. At this time Whitehouse was overworked and suffered severely from fever.[88] Dispatches from London offices urging the chief engineer that the pace of construction should be raised left him in a difficult spot and were anything but welcome. This was the view of Sir Frederick Jackson, who as one of the first British administrators in East Africa was stationed at Nairobi in 1901. According to Jackson the committee in London was

> far from fair towards Whitehouse. While up to his eyes in work and overburdened with local troubles and responsibilities, he was perpetually being pestered by cables urging him to push along, and get through to the lake, in order to save the committee from being pressed to 'get a move on' by the Foreign Office.[89]

By 1902 Whitehouse's relations with the committee in Westminster had deteriorated to the extent that he contemplated terminating his employment with the Uganda Railway.[90] He was, however, persuaded to see the project through the final stages and when Whitehouse and his family left for Britain in the spring of 1903 the railway was nearing completion.[91] The extent to which he had fallen out with the committee is revealed in his diary upon his arrival in London. In May he attended a meeting with consulting engineer Rendel in the office in Great George Street where he was told that he 'would never get another job'.[92] He later called several times at Rendel's but was unable to get further interviews.[93] He also met with the committee in the Foreign Office and recorded: 'At F.O. in the morning at Committee meeting, finished off with Uganda Railway. Had it out with O'C [O'Callaghan] in morning but no satisfaction'.[94] Whitehouse spent the following months in London in search of a new job. He considered taking on a position with Cape Colony Railways[95] but he was eventually appointed chief engineer to a mountain railway project in Argentina. He spent the next decades in Latin America and died in London in 1938.[96]

The ability to obtain the best positions in British colonies depended much on one's connections and 'standing' in Westminster. Although Whitehouse was in charge of railway operations on the ground in East Africa he was not at the top of the engineers' pecking order in Westminster – a position that was occupied by Alexander Rendel and other consulting engineers whose networks are analysed in the next chapter. Whitehouse was drawn to London but also pushed to new projects overseas in pursuit of more attractive career openings in South Africa, Mexico, India, East Africa and eventually in South America. In this regard Whitehouse's career path was not exceptional. Westminster was a location that pulled in engineers searching for higher positions in the British spheres of influence overseas and it was the place from where they were dispersed to all corners of the globe. This dynamic was integral to the workings of engineering imperialism and it was felt also by chief engineers such as Whitehouse.

The case of civil engineer Charles O. Burge sheds further light on how the dynamics worked between Westminster and the engineers who made a living working on colonial projects. Like Whitehouse, Burge spent his life and made his career working on railway projects overseas. In 1854–1904 he was employed on works in Britain, America, Asia, Africa and Australia with the construction of the Hawkesbury Bridge in New South Wales as a crowning achievement that earned him the KCB.[97] Burge's career resembled Whitehouse's in the sense that he shifted from one overseas project to the next with brief periods in Britain in between. In his autobiography, published after his retirement under the title *The Adventures of a Civil Engineer: Fifty Years on Five Continents*, Burge revealed more about how the dynamics between Westminster and the colonial regions were felt by engineers.

Burge was born in Ireland in 1835 where he also was trained as an engineer. He was employed by an Irish company that also took on jobs in England and Burge was therefore directed to Westminster, a location that he confirmed 'was the headquarters of engineering on account of its nearness to the parliament house'.[98] Burge first arrived at the time of the great crisis in domestic railway construction in 1866–7 and it was difficult for a young engineer to find employment. He therefore took up a position offered by an acquaintance on a railway project in the Decca Province in West-Central India where he spent the next few years. In the early 1870s Burge had finished with the project and returned to Britain where he was, however, only able to find employment on small-scale railway projects. He recalled that 'the engineering work which I now undertook was spasmodic and badly paid' and he noted that this was a general problem for engineers who chose to pursue imperial careers:

It is a curious fact that, in the civil engineering profession, if a young man once goes abroad for any considerable time he is forgotten and loses his status at home. He is generally dependent on higher members of the profession for employment. These are fully provided with assistants, who, having been able to tide over depressions, have remained with him or others. The wanderer, therefore, must either stay abroad or, if he comes home, must seek in London each time another post beyond these isles. Such one, therefore, rarely reaches eminence at the head-quarters of the profession, which London undoubtedly is. This is perhaps natural and of course there are high positions abroad, though neither in rank nor emolument vying with those at home.[99]

Penetrating the closed ranks of Westminster engineers was difficult and according to Burge long stays in the colonies did not make it easier. Burge – who in his own words also suffered from 'what the Germans call wander-lust, the true fons et origo of the British Empire'[100] – took on his next assignment in South Africa. Here he spent five years as a chief engineer to narrow-gauge railways in the Western Cape before he was 'again adrift on the world, bound for the centre of most engineering possibilities, London'.[101] Following another brief stay in London, the next stop for Burge was New South Wales where he was employed as resident engineer to the Hawkesbury Bridge (1886–7), the last link in the railway network that connected Adelaide, Sydney, Melbourne and Brisbane.[102] Burge stayed in Australia for seventeen years and became president of the Royal Society of New South Wales and lecturer in engineering at the University of Sydney.[103]

For Burge and Whitehouse Westminster was a stopover on the way to new projects overseas. Nothing indicates that theirs were exceptional cases. In obituaries in the *Proceedings of the Institution of Civil Engineers* of engineers whose professional careers spanned these decades, this way of moving between the metropole and the colonial frontiers in India, South America, Australia or Africa represented a common career pattern. In one sense there were as many individual reasons for these movements as there were engineers. Upon the completion of the Uganda Railway George Whitehouse was told by consulting engineer Alexander M. Rendel that he should never get another job. Whitehouse was, however, a railway engineer of wide experience and he was capable of securing a position in Argentina. Burge's career path was different. He described himself as 'a rolling stone' and explained that his personal engineering diasporas arose from a mix of 'wander-lust' and the fact that he was only able to get 'spasmodic and badly paid' employment in Britain. This was, in all likelihood, a mix of motivations that Burge shared with other engineers. By the last quarter of the nineteenth century the choice for many civil engineers was between working on minor projects in Britain or larger ones in the colonies.

Certainly, there was substantial outward movement among the civil engineers. Some migrated permanently and carved out lives in the colonies, for example through employment in colonial public works departments or with the railways of a colony. Others travelled more and were only based in a given region for short intervals, but many engineers developed 'imperial careers', a term introduced by Lambert and Lester as a designation for 'the men and women who dwelt for extended periods in one colony before moving on to dwell in others'.[104] 'Imperial careering' was a common pattern among engineers for a number of reasons. Large engineering projects often took years to complete but once construction had terminated fewer engineers were required and usually for positions in maintenance and administration. Engineers were therefore likely to leave for new projects in other areas. Here construction experiences gained in one colonial region could open doors in others and often colonial projects offered better opportunities than domestic work. There was, moreover, a tendency that 'imperial careering' was self-reinforcing because – as was also noted by Burge – longer periods in the colonies could work to the detriment of the ability to obtain higher positions in Westminster.

Conclusion

Civil engineers have largely been ignored or overlooked by scholars who have explored intersections of urban locations and imperialism. Engineers, however, warrant a place on the historical map of imperial London. In this period, British civil engineers, and in particular the top layers of the profession, clustered within just a few streets in Westminster, the political heart of the city. For engineers engaged in infrastructural projects the existence in Westminster of imperial and colonial governmental institutions contributed much to the attraction of the location. It was imperative to be present where the worlds of imperial politics, finance, information and engineering intersected.

This analysis has described the metropolitan engineering landscape that characterized a way of life and that was constitutive of the actions and negotiations that took place in it. It has, moreover, begun to explore how imperial impulses were remoulding the landscape – tangibly, for example, in the construction of the ICE headquarters inaugurated in Great George Street in 1913. The choice of construction materials and the meticulously selected imperial ornamentation for the building were material affirmations of the view that the engineering profession at the time was an imperial vocation and that it was expected to remain so in the years to come. Furthermore, the cases of the 'imperial careers' of Whitehouse and Burge reveal some of the ways in which this engineering centre was connected to the wider world. These engineers were in charge of large engineering projects in different colonial settings but they depended also on Westminster throughout their careers. In relation to engineers such as Burge and Whitehouse, a metaphor of Westminster as 'the heart of the engineering profession' is accurate; the location functioned as the centre in a system of contraction and ejection; engineers flowed to the metropole from different corners of the empire and were from there – sometimes voluntarily, sometimes less willingly – driven to the regions of the world where the British were carrying out their engineering projects.

3 ENGINEERING NETWORKS AND THE GREAT GEORGE STREET CLIQUE

Professional society is based on merit, but some acquire merit more easily than others.

Harold Perkin (1989)[1]

This chapter analyses the networks of the consulting engineers who dominated the market for British engineering projects in Africa in this period. It focuses on the elite segment of the profession. On those few engineers who were among – or who at least could hope to aspire to – what *Queen's Magazine* called the 'Upper Ten Thousand'; that is, the roughly 10,000 people in late-Victorian society who sustained an annual income of over £10,000.[2] The chapter demonstrates that the market for consultancy to British projects in Africa was controlled by tightly knit networks of consulting engineers based in Westminster and it analyses the ways in which the empire became an integral and constitutive part of the business platform of those select engineers. It identifies the key agents and analyses a number of shared characteristics that underpinned their networks. The strength of these networks originated from systems based on trust and patronage – elements that remained important in spite of the growing professionalization of the engineering profession that also characterized this period.

From their base in Westminster consulting engineers developed close ties with other groups with imperial interests. This chapter unravels how these connections were established and sustained over time. It argues that alliances between consulting engineers and other British imperial groups were central for the development of British imperialism in Africa in this period. Other agents were, however, critical of the influence exerted by London-based consulting engineers over infrastructural development in the colonies. The final section of the chapter analyses the roots and consequences of this criticism, in particular as it was voiced in relation to railway construction in Nigeria.

Networks, Trust and Patronage

The theme of networks has for a long time attracted the attention of historians of technology as well as of the British empire. In this chapter, the network approach refers to relations between human beings and thus to the concerns that have occupied the historians of empire who in recent years have explored the role of personal networks and patronage across Britain's imperial system.[3]

Human beings engage in social networks that take many shapes and forms.[4] They can be networks based on business and professional interests, on kinship and friendship, on political, social, ethnic or migratory relations. The historical significance of networks can be said in part to 'lie in their role as mechanisms consciously utilized by their members', for example in 'the transmission of information, or patronage, or money, through the personal connections a network encompassed'.[5] When analysing the functions of networks a useful distinction can be made between formal and informal networks. Formal networks are linked to institutions and organizations devoted to, for example, religious worship, cultural and scientific concerns, business and professions, politics or philanthropy. Informal networks are less tangible and are based, for instance, on apprenticeships, family relations, employment and partnerships. In reality informal and formal networks often overlap. Thus, within formal institutions smaller informal and often more tightly knit networks can exist and thrive.[6]

Networks have been subject to attention from historians who study the generation and transference of trust in social contexts.[7] They see networks as a main vehicle in the generation of *ascribed trust*, the kind of trust that arises from people's shared characteristics.[8] Indeed, the strength of a network is dependent on collective attributes among its members, such as gender, family, ethnicity, shared upbringing, education, professional occupation, beliefs and leisure activities. Such shared features can strengthen emotional bonds, produce feelings of solidarity, reduce the costs of control and monitoring, and enable people to gauge the honesty of those people whom they resemble most. There is much to be gained from participation in a network. A network gives the benefit of long-term relationships; it can ease the obtainment of status, enhance the reputation of individuals and facilitate gossip and the exchange of information. Close interaction in networks also lowers the risk of misunderstandings by permitting non-verbal communication, and makes it possible to estimate the long-term capabilities of potential partners.[9]

Patronage was an important element in the networks of consulting engineers in this period. Historians of science usually associate patronage with the early modern period exemplified by the case of Galileo Galilei.[10] The same is also true of historians of engineering who have explored how patronage was of vital importance for the identity of Renaissance engineers and for the technical solutions they devised.[11] Engineers were of course in a very different situation in the late nineteenth century. Specialization, the introduction of formal education and meritocratic structures – the gradual 'rise of professional society' which the engineering profession was part and parcel of – meant that engineers operated in a very different world.[12] Yet, the rise of professional society did not mean that patronage structures had disappeared in Victorian civil engineering. On the contrary, businesses passed from father to sons; apprenticeships could lead to lifelong employment and over time to partnerships between former master and pupil; entry to the formal institutions and trusted positions in the engineering profession was based on personal recommendations and connections.[13] The networks of consulting engineers were thus to a large degree based on what the imperial historian Zoë Laidlaw calls '*strong ties* of friendship and obligation'. Strong ties had the advantage that they increased agents' 'chances of manipulating a network to their advantage' and 'encouraged repeated and more determined action'. By contrast, connections between acquaintances – weak ties – required higher maintenance cost, could break more easily under stress and were much less agile when business was conducted and information exchanged over large distances – as was the case in an imperial context.[14]

The most important study of the workings of patronage systems and networks among consulting engineers in the Victorian period has been conducted by Porter and Clifton, whose work focused on the embankment of the Thames built in the 1860s and 1870s. Their study showed that positions and contracts for the massive embankment projects were negotiated among a small group of Westminster consulting engineers and their immediate business partners – to the extent that advocates 'of competing Thames Embankment schemes referred to the Great George Street Clique as a monopolistic network'.[15] The works along the Thames involved the circles of engineers whose imperial connections are explored in this chapter. That is, the network dominated the embankment projects carried out in their Westminster backyard, but the engineers of the Great George Street Clique also developed their networks to encompass projects carried out across the British empire.

Consulting Engineers and Africa

This section unravels the complex business and career patterns of the key Westminster consulting engineers with African commitments. It identifies shared characteristics of the consultants and pinpoints similarities in the platforms from which they became engaged with projects in Africa. A particular aspect brought out in the analysis is that connections established on the domestic scene and in other areas of the globe proved essential for entry to the market for African engineering projects.

John Fowler and Benjamin Baker – Egypt

John Fowler (1817–98) was the most renowned and well-reputed consulting engineer of the mid- and late Victorian era. In 1867 he became the youngest president of the ICE ever elected and he remained a towering figure in the engineering profession over the following decades. Fowler was above all associated with railways, including the first underground railways in London, and with the design of long-span bridges of which the Forth Railway Bridge in Scotland was the crowing engineering feat of a remarkable career.[16] Fowler, however, also had extensive business commitments throughout Britain's expanding imperial sphere, most notably in Egypt and New South Wales. Furthermore, some of the most influential consulting engineers to projects in tropical Africa after 1890 began their careers as apprentices and assistants in the office of Fowler.

Fowler began his career employed on railway projects in north England in the 1830s. His business expanded rapidly and in 1844 he moved from Sheffield to London to set up a consulting office (and family home) at 2 Queen Square Place in Westminster. He was thus among the first consulting engineers to establish himself near the Houses of Parliament and his work as an independent consulting engineer may be dated from that year. The office was a roomy house with a large garden situated by St James's Park between Buckingham Palace, the Palace of Westminster and the main building of the ICE. In 1873 the tall Queen Anne's Mansion was erected on the site and Fowler's office moved into this new building, where it remained until long after Fowler's retirement and death.[17]

Fowler's business grew from the beginning. It has been estimated that by the 1860s he was involved in 70–80 major schemes a year and that a professional career extending over 60 years must have amounted to over 1,000 jobs. The sheer scale of Fowler's business meant that he did not design all projects that the consultancy was involved with. Instead, the business made extensive use of assistants. So far more than 50 of these have been identified.[18] The most prominent of Fowler's assistants was Benjamin Baker, but William Shelford was another engineer who began his career in Fowler's office and later went on to found a consultancy with extensive commitments in Africa. Fowler's office was a place in which *strong ties* were forged, ties that were maintained and developed over decades.

While many of Fowler's business obligations were on domestic projects, he also worked in several foreign and colonial settings. He became particularly involved with railway construction in New South Wales. In 1856 Fowler's eldest sister married a talented engineer and former employee in Fowler's business, John Whitton, who migrated to become engineer-in-chief to the Government Railways of New South Wales. In 1860 John Fowler was hired as consulting and inspecting engineer in Britain to the rapidly growing railway system of that colony. This involved a number of tasks including the provision of purchase recommendations for locomotives, rolling stock and construction materials ordered from Britain as well as inspection responsibilities for materials and supplies before shipment to the Australian continent. Moreover, Fowler and his brother-in-law also collaborated closely on the design of a number of railway bridges, including three wrought iron bridges constructed in the 1860s over the river Nepean.[19] Fowler's business with New South Wales was initially small but partly due to his privileged connections with the colony's engineer-in-chief it grew substantial over the years. Between 1862 and 1887 Fowler was paid a staggering total of £56,968 in fees for his work for the Government Railways of New South Wales. By comparison, Whitton, by the time he retired as chief engineer to the railways in 1890, earned £1,800 annually, which was still enough to make him the second-highest salaried official in New South Wales (only the chief justice received a higher remuneration).[20]

Fowler's imperial platform was not confined to New South Wales. In 1870 he was appointed member of a commission to advise the government of India on the adoption of a narrower gauge for branch lines and in 1889 he visited the colony to give advice on the steep gradients on the Jhelum and Rawalpindi Frontier Railway.[21] In 1891–2 the Foreign Office requested an approximation from Fowler for the construction of the Uganda Railway. Fowler estimated that a railway could be constructed at £3,166 per mile. When the construction of the railway commenced a few years later the consulting engineer appointed was, however, Fowler's neighbourhood colleague, Alexander M. Rendel.[22]

On the African continent it was to be with Egypt and not Uganda that Fowler's name came to be associated. In 1869 he was hired as adviser by the Egyptian ruler, Khedive Ismail, who was embarking on a scheme to modernize the country by improving infrastructure, irrigation and industrial facilities.[23] Two years later Fowler was officially appointed consulting engineer to the Egyptian government under the title general engineering adviser – a position he held until 1879 when Ismail was deposed by the European powers.[24] During these years Fowler gave advice to the Egyptian government mainly in relation to three areas: the establishment of industrial plants for sugar-cane processing, the construction of the 'Soudan Railway' from Wadi Halfa to Khartoum and the development of irrigation systems in Lower Egypt.[25] The reports he submitted in the capacity of general engineering adviser show that his assistants made extensive surveys over the Egyptian lands during these years. In 1875, for example, a survey of the vast terrain between Wadi Halfa and Khartoum and further on into the Darfour province was carried out. This survey party was sent out 'with the view to obtain trustworthy information as to the practicability and cost of the proposed [Soudan] railway'. The survey party consisted of 8 English engineers, 4 Egyptian engineers, a doctor, 36 soldiers, 300–400 camels and 100–20 guards.[26] Like most of the projects reported on by Fowler in Egypt in this period the 'Soudan

Railway' never did materialize on account of the disarrayed financial situation of the Ismailian administration. With less than 100 miles of track laid, works were stopped and a compensation of £ 78,000 had to be paid to the contractors Messrs Appleby, with Fowler receiving fees as arbitrator.[27]

Ismail's modernization schemes produced neither an extensive infrastructural system nor an efficient agricultural and industrial sector in Egypt. They did, however, create an enormous debt – predominantly held by British investors – which was a major reason why the Khedive was removed from power.[28] Spearheaded by the British, the European powers replaced him with the puppet ruler Tawfiq and after the battle of Tel-el-Kebir in 1882 Egypt was *de facto* ruled by the British under a system already at the time known as a 'veiled protectorate'.[29]

When the British occupation of Egypt began in 1882 Fowler was well into his sixties and it was left increasingly in the hands of Benjamin Baker (1840–1907) to maintain the Egyptian connections of the consultancy. Baker had entered the office of Fowler as junior assistant in 1862 and remained with the company until Fowler's death in 1898. Under the wing of his patron Baker rose in the professional ranks and in 1875 he was taken on as full partner in Fowler's business. Baker was born in Ireland, educated at Cheltenham Grammar School and had been apprentice at Price & Fox's Ironworks in South Wales before joining Fowler's office in Westminster.[30] He had worked in Egypt with Fowler in connection with the surveys carried out for the khedival administration. Towards the end of his life he was also consulted on the scheme to construct a railway bridge across the Nile at Bulaq. Nearly three decades prior to this he designed the vessel that carried to London the ancient Egyptian obelisk known as Cleopatra's Needle.[31] Baker's most significant work in Egypt, however, related to the damming of the Nile at Aswan. He was consulting engineer to this project from 1894 until his death in 1907.

Baker's work in colonial settings was not confined to Egypt. From a relatively young age he established himself as a world expert on the strength of materials, bridge designs, railways, canals and irrigation projects. This reputation he developed and sustained through his connections with Fowler, the numerous projects he was affiliated with and an impressive list of technical publications on a wide range of topics including the strength of beams, bridge construction, urban railways and stress distribution in masonry dams.[32] As we shall see in later chapters Baker was also skilled at manoeuvring in the public sphere (Chapter 5) and at pulling political strings in the ICE where he was elected president in 1895–6 (Chapter 4).

Baker is the most lucid example of a consulting engineer who through his engineering skills, public reputation and personal connections became engaged in projects all over the globe and in particular throughout the expanding British empire. He did consulting work on numerous projects in Canada including the (failed) Chignecto Ship Railway across the isthmus that connects Nova Scotia with the American continent.[33] He was appointed consulting engineer to the Public Works Department of Cape Colony and in conjunction with Fowler he was consulted on the railway projects in New South Wales already referred to. After Fowler's death in 1898 he maintained the company office at 2 Queens Square Place, now in partnership with Arthur D. Hurtzig under the name Baker & Hurtzig. In 1905 he also established a partnership with Frederic Shelford (of whom more shortly) and through this connection advised on railway projects in Barbados, Cyprus and West Africa as consulting engineer to the office of the Crown Agents of the Colonial Office.[34]

Shelford & Son – West Africa

It was briefly noted above that one of Fowler's assistants, William Shelford (1834–1905), became one of the most important consulting engineers to projects in Africa in this period – mainly in the capacity of consultant to the Crown Agents. Unlike Benjamin Baker, however, Shelford did not build his imperial platform from a long career in Fowler's office but instead established an independent consultancy business on his own. Shelford was a member of council in the ICE and would have become president in 1902 had he not declined the nomination on account of his failing health.[35]

Shelford descended from a long line of clerics in Cambridge and was educated from Marlborough College.[36] After graduation he was taken on as a pupil with Messrs Gale, a Glasgow company specialized in waterworks. In the evenings young William attended mathematics and engineering classes at Glasgow University. By 1856 he had made such great progress (or so his daughter Anne Shelford assures us) that he was liberated from his apprenticeship and was advised to go to London. In London, he entered the office of Fowler where he stayed for four years. Shelford spent a few more years in the London office of consulting engineer T. F. Turner but in 1865 he engaged in a practice on his own account in Westminster, in partnership with Henry Robinson who later became professor of engineering at King's College. The company experienced troubled times from the beginning. Following the stock crises in 1867 ('Black Friday') Shelford was forced to let out the family house in London and move to smaller dwellings. He managed, however, to keep the business afloat and over the following years he and Robinson were consulted on domestic railway and waterworks, coal mining in Wales, gold mines in Nova Scotia and sulphur fields in Sicily. In 1875 the partnership with Robinson was dissolved and Shelford left the offices he had kept at Westminster Chambers and took on those at 35A Great George Street.[37]

It was from these offices that over the next three decades he directed a staff of assistants, wrote reports and supervised the works with which he was connected in most parts of the globe. The largest domestic project to which he was consultant was the Hull and Barnsley Railway (1880–5) but increasingly his business concentrated on work overseas. In 1878 he was appointed engineer to the Boston District tramway and he was consulted in connection with the Hudson Bay Railway in Canada. From the late 1880s he became involved with colonial projects in the tropics. In 1887 he prepared for the government of Ceylon and the Colonial Office a scheme for a narrow-gauge railway in Ceylon and the following year he was appointed consulting engineer to the Sungei–Ujong Railway in the Malay Peninsula. Together with George Barclay Bruce, a past president of the ICE, he furthermore proposed a highly ambitious extension of the South Indian railway system across the straits to connect India with Ceylon by running trains from Madras to Colombo. About this time his connections with South America also began. In 1890 Shelford was in the Argentine republic to report on a proposed railway from Rosario to Cordoba and to inspect the Valparaiso and Trans-Andine Railway – a mountain railway in the Andes. In 1892 he was appointed director of the Santa Fé and Cordoba Railway in Argentina.[38]

In the 1890s Shelford's business expanded into new regions including tropical Africa. Shelford's obituary in the *Proceedings of the Institution of Civil Engineers* stated that 'his name will probably be best remembered as Consulting Engineer to the Crown Agents for the Colonies, in which capacity he directed exploration surveys, and designed and built hundreds of miles of railways in Sierra Leone, Lagos, Gold Coast, and Nigeria'.[39] The first trajectory survey in West Africa carried out by his company was in Sierra Leone in 1893 and the section of the railway from Freetown to Songo Town opened in 1899. In the Colony of Lagos (later to become the Colony of Southern Nigeria) the first surveys were carried out in 1894 and the Lagos–Ibadan Railway was inaugurated in 1901. The Gold Coast surveys began in 1896 and the Tarkwa–Kumassi line opened for traffic in 1903.[40]

The widening of Shelford's activities overseas occurred when he was well into his fifties and in declining health. Operations on the ground were therefore left in the hands of assistants and his son Frederic Shelford (1871–1943), who obtained a degree in engineering from London University in 1891 and immediately entered the family business. He was initially placed in charge of a railway trajectory survey in Cornwall and thereafter became affiliated with the projects in West Africa. In 1899 he went into partnership with his father under the name Shelford & Son, and this company continued its operations until the death of William Shelford in 1905. Thereafter Frederic Shelford set up a new partnership with Benjamin Baker – his father's trusted 'old comrade' from Fowler's office and the council room of the ICE.[41] The consultancy kept the offices in 35A Great George Street, now under the name Baker & Shelford. Frederic Shelford stayed in the business of colonial engineering consulting after the death of Baker in 1907 – thereafter in partnership with another Westminster consultant, Robert Elliott-Cooper, who became president of the ICE in 1912. In addition to work in West Africa Shelford and Cooper also took on assignments in Kenya and Nyasaland.[42]

John Coode & Son – West Africa and South Africa

Coode & Son were consulting engineers to numerous colonial harbour projects in Africa in this period.[43] John Coode's (1816–92) way into the colonial consultancy business followed a pattern that by now is familiar: a successful career with domestic projects was followed by still more assignments overseas in the last quarter of the century. Coode did his apprenticeship with the harbour engineer James M. Rendel – father of Alexander M. Rendel. In the 1850s he established an independent consultancy practice in Westminster, where he was permanently resident from 1865. Coode was president of the ICE in 1889–91. He was consultant to major domestic works, most notably the construction of the colossal military harbour in Portland (1847–72) on the south coast of England. Coode, however, also developed a business with projects overseas.[44] From the 1850s his company was consulted in relation to proposed harbour projects by numerous colonial governments in the British empire including the Australian colonies and in Cape Colony and Natal. He was also consultant to the government in India and to harbour works in the Crown Colonies through the Crown Agents. In the latter capacity he was consulted on the large harbour works for Colombo in Ceylon and towards the end of his career for harbours on the West African coast – in Sierra Leone, Lagos and the Gold Coast respectively.[45] By this time his son had joined the consultancy to become partner in Coode & Son and took part in the first harbour surveys on the West African coast. In West Africa – and indeed in most other places – where Shelford & Son were employed as consulting engineers to colonial railway projects, Coode & Son provided consultancy services for the harbour works connected with the railways. Coode's grandson, Arthur Trevenen Coode, became partner in the firm in 1906 and designed numerous dredgers supplied to colonial governments through the Crown Agents.[46]

Douglas Fox & Partners: South Africa and Rhodesia

When consulted on African projects Fowler, Baker, the Shelfords and the Coodes were employed in the capacity as consultants to either the British or to colonial governments. In this regard Douglas Fox & Partners – consulting engineers with even more extensive commitments in Africa in this period – were in a different position. This company was not employed and paid by governmental bodies but by the British South Africa Company (BSAC), the chartered company that was founded in 1889.[47]

Douglas Fox & Partners was another family-run company with roots stretching back to the railway mania of the 1840s. It was founded by Charles Fox (1810–74) who is best known today for co-designing the 'Crystal Palace' for the 1851 World Exhibition (a design which earned him a knighthood). In 1856 Charles Fox's engineering and contracting business went into liquidation after a failed contract for a railway in Denmark, but the following year he established himself as a successful consulting engineer working from offices in Spring Gardens, Westminster.[48] In 1861 Charles Fox suffered an accident at a construction site from which he never fully recovered. As a consequence his young sons, Charles Douglas Fox (1840–1921) and Francis Fox (1844–1927), had to give up their planned studies at Cambridge and enter the business under the name Charles Fox & Sons. After the death of Charles Fox in 1874 the name changed to Douglas Fox & Partners.[49] The oldest son, Charles Douglas Fox, was elected president of the ICE in 1900.

From the beginning Charles Fox took up consultancy work on domestic projects as well as overseas. Early projects in the colonies included the design of the first railway bridges in Queensland and at the Cape of Good Hope as well as consultancy work in relation to narrow-gauge railways in Canada and India. Over the following decades the company's overseas activities expanded in size as well as in geographical scope. In the 1880s Douglas Fox & Partners' commitments included functioning as consultant engineers to the Central Argentine Railway Company, the South Indian Railway Company, the Southern Sao Paulo Railway Company and the Dorada (Colombia) Railway Company.[50]

After 1889 the company also operated extensively in southern Africa. In a period stretching more than three decades Douglas Fox & Partners were consulting engineers to thousands of miles of railways and several large bridges constructed in the territories administered by the BSAC – territories that at the height of the chartered company's rule covered an area nearly five times the size of Great Britain.[51] Douglas Fox & Partners held the position as consultants to the Shire Highlands (Nyasaland) Railway Company, the Rhodesia, Mashonaland and Beira Railway Company, the Benguella Railway Company and the Trans-Zambesia Railway Company – mainly subsidiary companies of the BSAC.[52] Thus Douglas Fox & Partners were consultants on all the railways extending north from Cape Colony and which in 1909 reached the copper mines of Katanga in the Belgian Congo approximately 1,500 miles north of the borders of Cape Colony. They were also consulting engineers to the lines that in 1902 efficiently linked up Rhodesia with Northern Mozambique and the Indian Ocean as well as to the line connecting the Rhodesian railway system westward with Angola and the Atlantic Ocean (the protracted Benguella Railway project of 1903–28).[53] This brief outline of the actives of Douglas Fox & Partners again exemplifies a general trend in the British engineering consultancy profession: it was companies that over time had generated professional experience as well as connections and contacts that increasingly became involved in large-scale engineering projects throughout the British empire – and by the end of the century also in the empire in the making in Africa.

For more than three decades Douglas Fox & Partners' extensive operations in southern Africa were managed by Charles Metcalfe (1853–1928). Born in India as the only child of Theophilus John Metcalfe, fifth baronet of the Bengal Civil Service, Metcalfe's social background was unusual for a late-Victorian civil engineer. Sir Charles, who succeeded his father as Baronet in 1883, was educated at Harrow School (1867) and at University College, Oxford (1874–7), graduating with second-class honours in law and history. Upon the completion of his Oxford education Metcalfe was articled to Douglas Fox & Partners where he remained for his entire career and in the mid-1890s became partner in the company. Metcalfe lived in southern Africa for most of his adult life and was elected member of council of the ICE in 1904.[54]

Shared Characteristics

There were several important shared characteristics among the consulting engineers who secured slices of African business from the last quarter of the nineteenth century. Firstly, all of the consultancies worked from offices in central Westminster. As described in Chapter 2, for engineers this location was much more than a mere business address. Westminster was integral to the identity of the consulting engineers who congregated in the central streets of the district. The area contained the venues for social and professional exchanges between engineers, politicians, clients and associate partners. Presence in Westminster conveyed power and status onto the dwellers. Most importantly, it was the ideal location for establishing and maintaining networks – and for renewing and redirecting them when required. It is therefore significant that the most successful engineering consultancies were established in Westminster in the early and mid-Victorian period. These companies had gained a foothold in Westminster working on projects in Britain. From this platform they later secured work on projects overseas, including tropical Africa. Parliament was of vital importance for the construction of domestic railways and consequently also for consulting engineers when offices were established in the area in the 1840s and 1850s. However, the consultants were increasingly hired by the imperial and colonial institutions that were also located in the same streets. Gradually, the ties and connections of Westminster consulting engineers became related to the empire.

Another shared characteristic was the way in which a career path followed a trajectory from apprenticeships, assistant positions and if exceptionally successful to partnerships. Patronage was not a thing of the past. This system of long-term affiliation and collaboration was clearly based on and further nursed 'strong ties of friendship and obligation'. By the last quarter of the century the path to a successful career as an imperial consulting engineer in most cases began with an apprenticeship with a well-established practitioner. This was a costly affair for a young person with ambition. Based on the scarce information available, W. J. Reader estimated that an apprenticeship with a top consultant in the 1870s required the payment of a premium of up to £500 directly to the consultant and a further outlay of £500 to cover all other expenses during the two years an apprenticeship normally lasted.[55] Well-reputed consultants were in a favourable situation to set the terms. When William Shelford arrived in London as a young man he worked for Fowler without salary for several months. This was after he had finished his apprenticeship with an engineer in Scotland.[56] For some younger engineers it was worth the investment. Upon the termination of the apprenticeship period it was occasionally possible to continue as assistant and from there rise further in the ranks.

Alternatively, engineers could seek employment with other independent consultants or set up their own practices. This was a dynamic process. Formal business partnerships were forged within a small circle and often between former apprentices or between junior and senior practitioners. Partnerships between Westminster engineers were even set up for individual projects when this was considered mutually beneficial – as for example when the experienced, former president of the ICE, George Barclay Bruce, in collaboration with the younger and well-connected William Shelford, submitted a proposal to the Colonial Office to link Ceylon and India via rail. This scheme never materialized but other and less imaginative projects sometimes did. The well-informed obituaries in the *Proceedings of the Institution of Civil Engineers* provide practically countless examples of Westminster engineers jointly putting forward proposals for specific projects or who, in partnership, were consulted by governmental bodies to particular engineering works in different regions of the empire. This system of advancement through apprenticeships, assistant positions, diverse and often uneven partnerships was nebulous. Younger engineers could spend years in the lower or middle ranks in Westminster or alternatively be forced to pursue engineering careers elsewhere. The cases of the wandering lives of Charles O. Burge and George Whitehouse discussed in the preceding chapter are illustrative examples of the latter strategy successfully employed.

Another common denominator in the engineering networks was the strong element of kinship that underpinned key connections. David Sunderland argues that kinship is a particularly strong force in networks as it is based on biology and ultimately on the desire to ensure the survival of a family's genetic inheritance.[57] We should in all likelihood not reduce the importance of kinship purely to biology, but family connections clearly cut a path to higher positions and colonial work that was much shorter than where such connections were absent.

In particular, it was common for sons to enter family businesses. This system provided better career opportunities for the young generation, and for the older it was a way to ensure the continuation of the family business and engineering dynasties they had established or in some cases inherited from their own fathers. This point was noted by W. J. Reader in relation to the leading engineers of the 1840s and 1850s and discussed more thoroughly by Porter and Clifton in relation to the dominant consulting engineers of the 1860s and 1870s.[58] The passing on of businesses and positions from father to son also continued during the 1890s and well into the next century.

There was a gradual change in the pattern, however, in that the younger generation of engineers usually had longer formal educations than the older and sometimes received a degree in engineering before they entered the family businesses. Yet, practical experience from the field of engineering works was still considered essential for the training of the future generation of consulting engineers. With respect to this the empire was also considered of great importance. It was practically standard procedure for the younger generation to go to the imperial outposts to learn the trade of the engineering consultant. The son of Francis Fox, Charles Beresford Fox, was stationed in Rhodesia in 1904–5 as resident engineer to the Victoria Falls Bridge.[59] Alexander M. Rendel's youngest son, Henry Wedgwood Rendel, was in East Africa in 1903 during the final stages of the construction of the Uganda Railway.[60] When Frederic Shelford joined the family business his father sent him to British Honduras to oversee the construction of an iron jetty. In 1899 the young Shelford was in the Gold Coast to lead a railway trajectory survey in the unmapped territory between the coast and the Ashanti capital, Kumasi. In 1903 he did another tour, this time to inspect the construction works then in progress in the West African region.[61] The grandson of John Coode, Arthur Trevenen Coode, was privately educated at Cambridge before joining the family business in 1903 to become partner only three years later. In 1908 he led an expedition sent out to charter the coastline off the Island of Lagos. The purpose of the expedition was to collect information for a planned deepening of Lagos Harbour that Coode & Son originally had designed a decade earlier.[62] John Fowler is missing from this list. None of his sons went into engineering.[63] He did, however, send his assistant and future successor, Benjamin Baker, to Egypt and South Africa.

The older generation of the Shelfords, Foxes, Coodes and Rendels had also worked on projects overseas in their younger years and clearly regarded imperial frontiers in Africa a proper training ground for their sons – the place to learn the trade and to build the character required of the future leaders of the profession. Frederic Shelford also consented to this point of view. In an article series he contributed to *Engineer* in 1908 he explicated this point:

I would give a few words of encouragement to any young man who may hesitate about leaving home. There is nothing in the world so good for a man, say, between twenty-five and thirty-five years of age, especially if unmarried, as a visit to distant countries, to learn tongues, to see new sights, and to experience strange adventures, such as he may readily encounter by throwing aside the restrictions of home life and taking up work abroad.[64]

The question of how engineering work in Africa appealed to different groups of engineers is important and complex. More is to be said about this in later chapters but here a specific point deserves to be highlighted. The biographies of the younger generation do show that formal education was increasingly becoming part of the background of consulting engineers in the last quarter of the nineteenth century. In this sense it confirms the received view that a growing formalization and professionalization of the engineering vocation was taking place at the time. Yet, we should be careful not to overemphasize the scope of this change. The older generation took pride in having won their fortune, fame and status from a background of little formal education and valued the trials of fieldwork in remote areas. Clearly, the ideals of formal, theoretical, scientific training of engineers that gained ground at this time co-existed alongside other deep-rooted traditions that stressed the importance of educating engineers through the practical trials of the colonial fields.

An essential characteristic shared by imperial consulting engineers with African connections was that they obtained high positions of trust in the ICE. The ICE was by far the most important formal platform of the Great George Street Clique. William Shelford and Charles Metcalfe served as members of council while John Fowler, John Coode, Benjamin Baker and Charles Douglas Fox were all elected presidents of the ICE in the course of their careers. This was not coincidental and must be seen as a reflection of a larger trend. It has already been demonstrated that in this period the council of the ICE was dominated by Westminster dwellers (see Table 2.1 above). Table 3.1 shows that consulting engineers with well-established Westminster businesses also dominated the presidency of the ICE during these decades. The table, moreover, shows that most of the consulting engineers who were elected president had widespread business commitments in the colonial world. The extent to which the leading consultants depended on a colonial and imperial platform differed. Yet, few had no imperial links and many more had a lot. There was a clear convergence between on the one hand high positions in the ICE and on the other extensive affiliations with imperial projects. The two aspects were in fact mutually reinforcing: a high position in the ICE enhanced the ability to obtain imperial projects while imperial projects provided a path to the highest positions in the ICE.

Table 3.1: Addresses and Colonial Connections of Presidents of the Institution of Civil Engineers, 1880–1914[65]

Year	Name	Occupation and Main Field	Address	Colonial Connections	Crown Agents Assignments
1862–3	John Hawkshaw	Consulting engineer, railways/docks	Westminster (est. 1850)	India, South America, Egypt	x
1866–7	John Fowler	Consulting engineer, railways/bridges	Westminster (est. 1844)	Egypt, New South Wales	x
1880	William Henry Barlow (1812–1902)	Consulting engineer, railway/bridges	Westminster (est. 1857)	India, New South Wales	x
1881	James Abernethy (1814–96)	Consulting engineer, harbours	Westminster (est. 1854)	Egypt	
1882	William Armstrong (1810–1900)	Arms manufacturer	Newcastle		
1883	James Brunlees (1816–92)	Consulting engineer, railway/bridges	Westminster (est. 1856)	Brazil, Uruguay	
1884	Joseph W. Bazalgette (1819–91)	Consulting engineer, sanitation/waterworks	Westminster (est. 1842)		
1885–6	Frederick Bramwell (1818–1903)	Consulting engineer, mechanical/naval	Westminster (est. 1853)		
1887	Edward Woods (1814–1903)	Consulting engineer, railways/water	Westminster (est. 1853)	Argentina, Chile, Peru, Mexico	x
1888–9	George Barclay Bruce (1821–1908)	Consulting engineer, railway/bridges	Westminster (est. 1856)	India, Canada, Cape Colony	x
1890–1	John Coode (1816–92)	Consulting engineer, harbours	Westminster (est. 1844)	Cape Colony, Natal, New Zealand	x
1892	George Berkley (1821–93)	Consulting engineer, railways/bridges	Westminster (est. 1849)	India, Cape Colony, Natal, Argentina	x
1893	Harrison Hayter (1825–98)	Consulting engineer, railways/harbours	Westminster (1870, partner J. Hawkshaw)	India, Jamaica, Mauritius	x
1894	Alfred Giles (1816–95)	Engineer, railways/docks	Southampton	Canada	
1895	Robert Rawlinson (1810–98)	Consulting engineer/army engineer	London		
1896	Benjamin Baker (1840–1907)	Consulting engineer, railways/bridges	Westminster (1875, partner J. Fowler)	New South Wales, Cape, Egypt	x
1897–8	John Wolfe Barry (1836–1918)	Consulting engineer, railways/harbours	Westminster (est. 1867)	India, Natal, Shanghai	x

Year	Name	Occupation and Main Field	Address	Colonial Connections	Crown Agents Assignments
1899	William Henry Preece (1834–1913)	Consulting engineer, telegraphy/electricity	Westminster (est. 1899)	Nigeria, India	x
1900	Charles Douglas Fox (1840–1921)	Consulting engineer, railways/bridges/mining	Westminster and City (est. 1857, by father)	Argentina, Columbia, Rhodesia, South Africa	
1901	James Mansergh (1834–1905)	Consulting engineer, sanitation/water	Westminster (est. 1866.)	Colony of Victoria, Ceylon	x
1902	Charles Hawksley (1839–1917)	Consulting engineer	Westminster (1866, father's business)	Brazil, India	x
1903	John C. Hawkshaw (1841–1921)	Consulting engineer, harbours/railways	Westminster (1870, father's business)	Egypt, South America, India	x
1904	William White (1832–1913)	Consulting engineer, railways/bridges	Westminster (est. 1900)	India (stationed there until 1900)	
1905	Guildford Molesworth (1828–1925)	Consulting engineer/railway manager	London	Ceylon, India, Uganda (stationed in India until 1888)	
1906	Alexander R. Binnie (1839–1917)	Consulting engineer, waterworks/railways	Westminster (est. 1900)	India, Canada, Malta	x
1907	Aleander Kennedy (1847–1928)	University professor/consulting engineer, electricity	Westminster (est. 1889)		
1908	William Matthews (1844–1922)	Consulting engineer, harbours	Westminster (1892, partner in Coode & Son)	Cape, Hong Kong, Cyprus	x
1909–10	James Charles Inglis (1851–1911)	Railway engineer/manager	Rottingdean		
1911	Alexander Siemens (1847–1928)	Manager/consulting engineer, electricity	London (1883, director Messrs Siemens)		
1912	William C. Unwin (1838–1933)	University professor	Cambridge		
1913	Robert Elliott-Cooper (1845–1942)	Consulting engineer, railways/bridges	Westminster (est. 1876)	Gold Coast, Lagos, India	x
1914	Anthony G. Lyster (1852–1920)	Dock engineer	Liverpool		

It must be emphasized that the shared characteristics were not endemic to the consultants who worked in Africa. From Table 3.1 the case of the Hawkshaw family may be taken as an example of a successful imperial engineering dynasty that embodied the identified characteristics without expanding its operations into the African continent. The beginning of this engineering dynasty dates to 1850 when John Hawkshaw established a highly successful engineering consultancy in Great George Street.[66] Hawkshaw was elected President of the ICE in 1862. In 1870 his son, John C. Hawkshaw, and his former chief assistant, Harrison Hayter, joined as Partners and both eventually became ICE presidents, Hayter in 1892 and John C. in 1902. The company had extensive commitments in India as consultants to railway, bridge and dock projects. Their business in South America, in particular Argentina, was even more substantial, leading to the establishment of a permanent office in Buenos Aires in 1872, making the Hawkshaws the first British consulting engineers to establish a branch office outside Britain.[67] Moreover, the company was involved with railway construction in Mauritius, Jamaica, British Guiana and Barbados in their capacity as consulting engineers to the Crown Agents.[68] Partnerships were also established for overseas projects with several other past and future Presidents of the ICE, including John Henry Barlow and John Wolfe Barry – the latter began his long career as assistant to John Hawkshaw senior.[69] However, in spite of the substantial imperial commitments the company did not work in mainland Africa, except for Egypt where John Hawkshaw in the early 1860s gave favourable estimates for the construction of the Suez Canal that was later designed by French engineers. Thus, the case of the Hawkshaws shows that it was possible to belong to the Great George Street network and develop an extensive imperial consultancy business without entering the burgeoning sub-Saharan engineering field.

A number of shared characteristics thus underpinned the platform of a small, elitist group of consulting engineers who dominated the market for consultancy to projects in Africa – and indeed across many parts of the British empire. These characteristics included: early establishment of a consultancy practice in the heart of Westminster; apprenticeships and partnerships based on kinship, long-term connections and 'strong ties of friendship and obligation'; obtainment of positions of trust in the ICE and a history of affiliation with imperial projects outside Africa. These were basic ingredients in the recipe for entry to the market for engineering consultancy in Africa. There was, however, room for much contingency as to exactly which consultants were hired when, where and on what terms. This depended primarily on how individual consultants were connected with other groups with imperial interest. It is to these connections the chapter now turns.

Imperial Connections

John Fowler's connections with Egypt began during a leisurely trip to the country in 1869 where he travelled in the company of distinguished friends including the Duke of Sunderland and Professor Richard Owen. At the time Egypt was a favourite destination for the British upper classes and the band of travellers visited the newly inaugurated Suez Canal (escorted by Ferdinand De Lesseps), and were invited to accompany the Prince and Princess of Wales on a journey down the Nile. The trip took a professional turn when Fowler, through his influential friends, was given an audience with Khedive Ismail, who hired the prominent engineer to advise him on the schemes to modernize Egypt by means of industry and infrastructure.[70] According to Fowler's well-informed contemporary biographer, Egypt under Ismail's reign was a playground where 'distinguished English engineers – Armstrongs, and Rendels, and Bramwells – flit across the scene' and submitted to the khedival administration reports suggesting investments in industrial plants, railways, bridges, steam navigation and irrigation.[71] However, towards the end of Ismail's reign even consulting engineers like Fowler and Baker (who had become partner in Fowler's business in 1875) had trouble getting their fees from the indebted Egyptian Ministry of Finance.[72] Fowler's and Baker's connection with Egypt thus began as advisers to an independent ruler.[73] This changed, however, when Ismail was removed from power and Egypt became a 'veiled protectorate' controlled by the British. Imperial expansion accelerated and when the political tables turned, the consulting engineers' Egyptian field shifted from foreign to colonial.

In the first decade after the occupation this change in fact worked to the detriment of the consultants' opportunities in the region as the new British administration in Egypt allocated substantial resources to the public works department of the 'veiled protectorate'. This public works department was dominated by highly qualified Royal Engineers who transferred from India and who saw little need for consulting engineers from London; having gained their experience on large-scale irrigation projects on the Indian rivers they set about developing Egyptian irrigation systems without relying on experts with offices by the Thames.[74] This was particularly evident in connection with extensive and expensive repair works that in the 1880s were carried out on two irrigation barrages at the head of the Nile delta above Cairo. The barrages had originally been completed in 1863 but their hastily constructed foundations were insecure and suffered from wash-outs that made the structures incapable of holding back the Nile. During his time as Ismail's general engineering adviser, Fowler had suggested sinking a wall of concrete below the river bed to prevent water from washing out mud under the foundations. Fowler's approach was explicitly rejected by the engineers in public works department as costly and impractical.[75] Instead of deepening the foundations the Royal Engineers devised a project in which the structures were extended horizontally, downstream and upstream, thereby compelling the water to travel so far that its velocity would be checked. This was a method the engineers had successfully employed on rivers in southern India and which also proved efficient in Egypt.[76]

Thus, while imperial expansion usually worked to the advantage of Westminster consulting engineers the opposite was initially the case for Fowler and Baker in Egypt. However, when the immense scheme to dam the Nile at Aswan was instigated in the early 1890s Baker was hired as consulting engineer to the project. Baker's relations with the British administrators and engineers in Egypt are analysed in closer detail in Chapter 6, but in this context it should be noted that the consulting engineer's connections in London proved useful when finance for the Aswan dam was secured. Initially, the British rulers in Egypt failed to raise the capital needed for the dam as the government in London refused to grant or loan the £2 million required. Private financing was only secured in 1899 when Baker procured the contact between Lord Cromer, the London banker Ernest Cassel and the contractor John Aird & Son – a long-time business associate of Baker's and next-door neighbour in Westminster – who took on a financially creative contract for the project in which the contractor received payment in instalments over 30 years.[77] Just like Fowler had been during the reign of Ismail, Baker was centrally placed in the circles that were capable of setting in motion far-reaching developments in Egypt.

The entry of Douglas Fox & Partners into the African market was even more closely connected with Britain's imperial expansion and consolidation on the African continent. Indeed, it was entangled with some of the formative events in the political history of southern Africa in this period, including the discoveries of the world's richest gold deposits on the Witwatersrand (1886), the founding of the BSAC (1889) and of the colony of Rhodesia (1895). These developments are well described in the literature but it is important to unravel how Douglas Fox & Partners secured a position in Africa because it allows a rare glimpse into how the alliances of engineering and imperialism were forged and developed over time.

Douglas Fox & Partners' connections with South African engineering projects began in London. In 1888 the consultants were employed by the London-based Exploring Company to carry out a survey for a railway through Bechuanaland – since 1885 a British protectorate immediately north of the Cape Colony. The Exploring Company was founded by a group of capitalists who, lured by the fortunes made on the South African goldfields discovered two years previously, aimed to find and develop mineral resources in Bechuanaland and territories further to the north. The Exploring Company had therefore made an arrangement with the Colonial Office that the company would receive mineral concessions in the region in exchange for constructing a railway through the impoverished Bechuanaland protectorate.[78] Douglas Fox & Partners on behalf of the Exploring Company negotiated a preliminary contract for the railway with the Westminster contractors Lucas & Aird.[79] Charles Metcalfe, at this stage a young assistant engineer in Douglas Fox & Partners, was sent to South Africa to carry out the trajectory survey.[80]

In South Africa, however, things developed fast. Business tycoon and imperialist Cecil Rhodes also had his eye on the territories north of the South African colonies and used his economic and political influence in the region to prevent the Exploring Company from proceeding with the railway project. Instead, Rhodes forced through an amalgamation between the Exploring Company and the group of South African capitalists he spearheaded. Charles Metcalfe, who knew Rhodes from student days in Oxford, procured the contact between the two groups.[81] The following year the amalgamation of the London-based and South African capitalists formed the BSAC and secured from the British government a Royal Charter that gave the company a free hand to administer an undefined territory north of the South African colonies. Significantly for the consulting engineers, in the Royal Charter the Colonial Office insisted that the obligation to construct the railway through Bechuanaland fell on the BSAC.[82] Hence, Douglas Fox & Partners and Metcalfe designed the railway but now as consultants to a company with far-reaching economic and political ambitions in the entire South and Central African region. These ambitions were pursued over the years that followed as the BSAC violently subjugated the vast territory which in 1895 was named Rhodesia (after Rhodes).[83]

The BSAC and Cecil Rhodes in particular proved not to be very eager railway builders when money for projects had to be provided from the strained finances of the chartered company.[84] However, in the longer term the connection with the BSAC proved very favourable for Douglas Fox & Partners. As Jon Lunn has shown in detail, by the mid-1890s a mutually beneficial financial alliance was established between the BSAC, the railway contractors Pauling & Co. and the merchant bankers the House of d'Erlanger. The relations between the parties were complex but the basic model worked this way: d'Erlanger became the most important external investor in the BSAC's financially stretched railway schemes. At the same time d'Erlanger partially owned the contractors Pauling & Co. and consequently made a profit whenever Pauling & Co. did. In return for d'Erlanger railway investments the BSAC used its governmental powers to give Pauling & Co. monopoly of the construction of railway lines in the region. This basic model could be manipulated in different ways and railway construction was pushed forward mainly to service the burgeoning mining industries in Rhodesia and on the borders of the Congo Free State.[85] More than 2,500 miles were built over the next 30 years, with Douglas Fox & Partners and Metcalfe functioning as consultants on all of them. The consulting engineers had thus obtained a position in the 'charmed circle' around Rhodes and the BSAC.[86]

The highly creative financial set-ups behind the railways built under the auspices of the BSAC were negotiated within a small and long-lasting network of people that also included the consulting engineers. They were trusted insiders who besides being experienced railway builders also knew how the system functioned. It deserves mentioning that the few historians who in passing have commented on Douglas Fox & Partners' connections with the BSAC have invoked Charles Metcalfe's personal friendship with Rhodes in order to explain the position of the consulting engineers within the 'charmed circle'.[87] The evidence for this is circumstantial and sentimental. Rhodes died in 1902 and the collaboration between the BSAC and Douglas Fox & Partners continued for more than 20 years thereafter. What ultimately kept the alliance going was that it was advantageous and lucrative for all parties involved.

William Shelford's African connections were established through the Office of the Crown Agents for the Colonies – a quasi-governmental body attached to the Colonial Office. The Office of the Crown Agents was an essential component in the administration of engineering projects in British Crown Colonies in Africa and elsewhere. From headquarters in Millbank in Westminster the Crown Agents organized the supply and transport of manufactured goods to Crown Colonies, supervised the expenditure of capital, issued loans and took on a wide range of other tasks including the establishment of infrastructure in the colonies. Through this host of functions the Crown Agents constituted a crucial link in a trilateral relationship between the Colonial Office, Crown Colony administrations and the suppliers of goods and services for the colonies – including the services of consulting engineers. There were never more than three agents but the office employed a group of assistants that by 1914 numbered 468.[88]

During the 1880s William Shelford's business expanded to projects in areas that from the vantage point of London were located in the most remote regions of the globe. In 1893 he carried out what was probably his first assignment for the Crown Agents when he was adviser on the design of engine equipment for railways in Malta. Around the same time the Crown Agents also hired Shelford as consultant to a railway line between Belize and El Cazo in British Honduras and to design iron jetties for the Island of Dominica in the West Indies. When the Colonial Office as part of Joseph Chamberlain's programme of constructive imperialism in the 1890s contemplated equipping the West African Crown Colonies of Lagos, Sierra Leone and the Gold Coast with railways, the Crown Agents also employed Shelford as consultant to the projected lines. In 1898 he was officially titled consulting engineer to the Colonial Government Railways of West Africa.[89]

William Shelford enjoyed the benefits of kinship connections with the Crown Agents. The agent in charge of engineering, Sir Maurice Cameron, was a close friend of Shelford's brother, Thomas Shelford.[90] In addition, in 1899 William's son Frederic married Mildred Ommanney, the second daughter of senior agent Sir Montagu Ommanney. The father-and-son partnership in Shelford & Son was thus established in the same year that *Queen's Magazine* announced that an 'interesting marriage had been arranged' between the two upper-class families.[91] In the case of Shelford & Son 'strong ties' of kinship underpinned the bonds in the consultancy as well as their imperial connections. These ties served the Shelfords well. Over the following years, the company enjoyed the full support of the Crown Agents and of the Colonial Office where Frederic Shelford's father-in-law, Sir Ommanney, was elevated to the high position of permanent under-secretary in 1901.[92]

In his thorough studies of the Crown Agents David Sunderland has examined in detail the connections between the Crown Agents and Shelford & Son. He has shown that it was a scheming relationship characterized by outright nepotism, in particular after Frederic Shelford joined the family business and took over from his ageing and feeble father. The projected railways in West Africa to which Shelford was consultant ran into constant delays and were overtly expensive compared with both pre-construction estimates and equivalent French and German lines constructed in the region. The finished lines were widely known to be of an extremely poor quality and became notorious for their excessive gradients and curves and for their defective drainage, culverts and ballasting. Indeed, several contemporaries independently of each other rated Frederic Shelford's engineering skills to be very poor.[93] Furthermore, for the West African railways a so-called 'departmental system of railway construction' was adopted which meant that most of the work was not put out to contractors. Instead, Shelford & Son was *de facto* contractor as well as consultant to the projects. This effectively undermined the tender process and enabled Frederic Shelford to favour specific manufacturers. Purchasing recommendations by the consulting engineers went unmonitored until 1909. Moreover, in addition to fees for pre-construction work and for supervision during construction in West Africa the consulting engineers against normal procedure also received a fixed salary and had expenses covered during periods of construction. As there were no contractual time limits and penalty clauses, Shelford, in order to maximize and prolong his remuneration, deliberately slowed the construction processes by introducing very little machinery to the African field.[94] There was much to criticize in the partnership between the Crown Agents and Shelford & Son, and as we shall see in the next section of this chapter a number of contemporaries did criticize it. However, Frederic Shelford's privileged connections with the Crown Agents enabled him to sustain a position and an income that in the first decade of the twentieth century secured him the luxuries and comforts of a life in the highest circles of British society.[95]

The case of Frederic Shelford was extreme but generally the Crown Agents were important clients and business partners to a number of consulting engineers who belonged to the inner circle of the Great George Street Clique. Indeed, when selecting companies for consultancy work the Crown Agents favoured expensive, well-established Westminster businesses from the top layers of the profession. An indication of this may be obtained from the fact that at least 14 of the 30 engineers who served as presidents of the ICE between 1880 and 1914 carried out consultancy work for the Crown Agents in the course of their careers. The figure may even be higher since minor assignments may have passed unnoticed in obituaries and other sources. The positive evidence does, however, clearly establish that the Great George Street Clique and the Crown Agents collaborated closely.[96]

Sunderland has persuasively argued that the Crown Agents, in order to ensure high-quality services, chose top consulting engineers for whom the cost of poor performance of their duties would be high, leading to the loss of their commercially valuable high status. From the point of view of the consulting engineer, working for the Crown Agents was also attractive for a number of reasons. Cost was not a primary concern for the Crown Agents and consulting engineers received handsome fees for the services they provided, but assignments in Crown Colonies entailed other advantages in addition to pecuniary incentives. Affiliations with the Crown Agents were usually long term. They sometimes lasted for decades, passing from one generation of engineers to the next, as in the case of the Hawkshaws and the Coodes. Another case in point in this respect was William Preece, consulting engineer in the field of electricity and president of the ICE in 1899. According to Preece's biographer, the Crown Agents were his most important clients. Preece's papers (which now unfortunately have been lost) contained over 1,000 communications with the Crown Agents comprising specifications, recommendations on tenders, contractors and equipment for telegraphs lines, underground and submarine cables, tools and clocks for railway signalling relating to assignments in Australia, the Caribbean, the Far East and in Africa, where Preece and his assistants were consultants to a number of projects in West Africa and South Africa.[97] One of these assistants was the young James Sivewright who during an inspection tour to Cape Colony in 1877 was convinced to remain and take on the position as general manager of telegraphy in the South African Colonies. As a political bridge-builder between Cape Colony and Transvaal he became one of the most influential statesmen in South Africa during the 1890s.[98]

Preece's connections with the Crown Agents were also passed on to the next generation. In 1899 he established a consultancy in Queen Anne's Gate in Westminster with his sons, Llewellyn and Arthur, as partners. They continued and developed the connections with the Crown Agents in the next century and carried out consultancy work in several colonial settings including the Malay Straits, Ceylon, Trinidad and British Guiana. In 1932 Arthur Preece was knighted for his services to the Colonial Office.[99] His father had been knighted in 1899.

In the case of the Preeces, long affiliation with the Crown Agents prompted knighthoods. This should be seen as part of another set of attractions. Honours were a mark of social standing that after 1850 also came within the reach of civil engineers. Indeed, as Macleod has shown the successful execution of infrastructural projects in the colonies was particularly likely to release the knighthoods that were awarded to prominent late-Victorian and Edwardian civil engineers.[100] For engineers with gentlemanly aspirations there was much social and professional status to be harvested from Crown Colonial assignments. Projects in these areas were prestigious in wider society as well as within the engineering profession. The fact that many engineers were consultants to the Crown Agents prior to their nomination for the presidency of the ICE further underscores this connection. Moreover, Crown Colonial projects were extensively covered by the press and publicity was an important attraction for consulting engineers who (as will be discussed in detail in Chapter 5) were not allowed to advertise their services through commercial channels.

Colonial Critics

Imperial connections and the engineering projects they generated were central to the businesses, societal aspirations and profession identities of the consulting engineers of the Great George Street Clique. Many contemporaries, however, criticized the extensive use of the metropolitan engineers and considered them redundant or directly harmful to infrastructural development throughout the British empire. This section examines the nature and consequence of the criticisms levelled by British subjects based in African colonies against what they saw as a monopolistic sway exercised by the network of Westminster consulting engineers.

Critical voices were in particular raised over the close collaboration between the select group of London consultants and the Crown Agents. One vocal group of critics was constituted by British manufacturers stationed in Africa who protested that the system favoured manufacturers based in south England. This complaint was regularly aired in the engineering press, and particularly frequent in the 'Letters to the Editor' sections of *African Engineering*. In 1905 one manufacturer stated that

> The effect of the Crown-Agent System is to delay and complicate business, to restrict the number of firms asked to tender, and though it may be profitable enough to a few favourite firms, it is dead against the interest of our manufacturers as a class.[101]

The critic was supported by another manufacturer who specifically insisted that it was when specifications 'get into the hands of the consulting engineer in London that the manufacturers' troubles commence'. He stated a list of complaints which gives a rare insight into the power wielded by consulting engineers through their tender specifications and purchase recommendations:

> How can he [the manufacturer] compete as to price, when he is obliged to alter his designs, entailing new drawings and patterns; when he is compelled to provide six set of photographs and three or four sets of handmade drawings for every machine he supplies; when he has to furnish bars for analytical and tensile tests of materials; and in addition pay a fee to the inspecting engineer for making such tests; when he has to purchase his materials from certain firms whose names are specified, instead of procuring in the open market? When he has to pack his goods in cases in many instances ridiculously over-strong for their contents? The subject is not a new one, and it crops up periodically in the engineering papers when some unfortunate contractor is goaded into making a protest, which is always in vain, as the consulting engineer invariably declines to get drawn into any controversy, and persistently ignores any attempts to induce them to depart from their hide-bound and time-honoured methods.[102]

Consulting engineers were placed in a position that enabled them to shape specifications to the smallest point. In the details of specifications they could influence profoundly who would be able to compete for contracts. The editorial columns of *African Engineering* backed the readers who complained about the methods employed by London consulting engineers. In 1909 a leader on 'the general consultant' asserted:

> We maintain that it is ridiculous at the present day to pass a great variety of work through one man, for the simple reason that there is not, and never will be, one man who can deal with these matters efficiently. And yet there are hundreds of specialist engineers in London, who have a complete knowledge in their own particular branch who would make thoroughly good consultants. Many of these men have succeeded, but many more are struggling for a livelihood, and no end of them go to the wall in the process because they have not the name and the influence to bring themselves in touch with the men who might wish to employ them.[103]

Such criticism of the Westminster consulting engineers was widespread. However, manufacturers and editors without 'the name and the influence' could do little to change the system in any significant way.

This does not mean, however, that consulting engineers were insuperable but only that it took more powerful groups than minor Africa-based manufacturers to interrupt their imperial connections. In this respect political administrators in colonies and engineers employed in colonial public works departments could muster a more forceful opposition. We have already seen that the Royal Engineers in the Egyptian public works department were in a position that enabled them to carry out the major repair works on the irrigation barrages in the Nile delta without the use of consultants and along lines different from those originally proposed by John Fowler. In Crown Colonies the opposition to the Westminster consulting engineers was even more noticeable. Indeed, Crown Colonial administrators and engineers regularly complained that consultants in Britain were paid too much for their services, that they were unfamiliar with local conditions in relation to labour and climate, and that the reliance on London-based engineers impeded the development of engineering expertise locally in the colonial public works departments. A closer examination of railway projects in the Crown Colonies in West Africa sheds further light on the nature and consequences of such criticisms. The West African case, moreover, warrants particularly attention because the colonial critics succeeded in ousting the consultants, Shelford & Son, from the lucrative projects in spite of the consultancy's official status as consulting engineer to the Colonial Government Railways of West Africa and, despite the kinship-based 'strong ties of friendship and obligation' that Shelford & Son enjoyed with the Crown Agents and the Colonial Office.

The first signs of problems for Shelford & Son's position in West Africa occurred in Northern Nigeria, a protectorate established in 1900 after the British government had withdrawn the Royal Charter from the financially and politically weak Royal Niger Company.[104] The appointed high commissioner for Northern Nigeria, Frederick Lugard, was an adamant critic of the Crown Agent system in general and Frederic Shelford's work in particular. In 1901 Lugard, despite expressed opposition from the Crown Agents and Shelford, managed to obtain from the Colonial Office a permission to construct a 22-mile narrow-gauge railway to link the new administrative headquarters of the colony at Zungeru with the Kaduna River.[105] This short railway was the first in the West African Crown Colonies to be built 'locally', that is without the use of London consultants. Encouraged by his success Lugard then proposed a larger locally constructed line from Baro on the north bank of the Niger to Kano, the most populous trade centre of the colony located 350 miles further north. Owing to lack of funds this more ambitious scheme only got under way in 1906 when the project received backing from the parliamentary secretary of state for the colonies, Winston Churchill, another arch-opponent of the Crown Agent system who saw this project as a welcome opportunity to test on a larger scale the system of local construction against that based on Crown Agents and London consulting engineers.[106]

In Nigeria the railway builders conducted a survey of the terrain to replace a rudimentary survey carried out by Shelford & Son two years previously. By the time the new survey was completed Lugard had been replaced as high commissioner by a former Royal Engineer, Percy Girouard, who continued the efforts to eject the London consultants from the colony's railway projects. An expert in colonial railway construction with wide experience from Sudan, Egypt and

South Africa, Girouard was in a favourable position to do so.[107] The projected Baro–Kano railway was upgraded from narrow gauge to the standard 3'6" gauge to make it compatible with the railway system in neighbouring Southern Nigeria and John Eaglesome of the protectorate's public works department was placed in charge of construction. Eaglesome, who was a full member of the ICE, had been employed in public works departments in India for ten years before transferring to Northern Nigeria to take on the position as director of public works.[108] According to Girouard – who confidentially kept Lugard informed on the progress of the railway – Eaglesome was 'a far more capable man than Shelford, and moreover knows the country, which the former has never visited'.[109]

The construction of the Baro–Kano railway began in January 1908 and without the use of consultants it broke in several respects with the construction system previously adopted elsewhere in the British Crown Colonies in West Africa. Speed and cost of construction became a major concern. All administrative resources that could be spared were allocated to the railway. Road works carried out at administrative headquarters were brought to a halt in order to secure additional manpower and funds. Staff from the political and military administration was transferred to the public works department and from there to the railhead to supervise construction gangs, to assist with earthworks under the guidance of the public works department engineers and even to fill positions in cases of sickness among the engineering staff.[110] In a final move to secure high construction pace, Girouard prevailed upon the War Office to send out a contingent of 33 Royal Engineers to assist the public works department in laying the track.[111] According to Girouard the change to local construction also had positive effect on the *esprit de corps*. He informed Lugard that 'here all of us are interested in the work and call it "our railway": on the other side [southern Nigeria] it is merely a Shelford extension' and he triumphantly confided that 'I sincerely trust that we have given the old system its death blow'.[112]

Owing to these combined measures the pace of construction was kept high but at the price of significantly lowered technical standards. A. L. Seymour, a young engineer employed as assistant on the railway, later characterized the Baro–Kano railway as 'a pioneer type line' and recalled that instructions were given to proceed as quickly as possible, leaving it to posterity to 'strengthen the whole structure if and when conditions required it'.[113] Track laying, for example, did not wait for the completion of permanent bridges. Instead rivers carrying little water in the dry season were filled with an earth embankment under which galvanized iron pipes were laid. Heavy and prolonged rain, however, occasionally overtopped the embankments and carried them downstream along with the pipes, leaving only the track behind. In cases of such washouts the whole work had to be reconstructed. The procedure did, however, allow track-laying gangs to continue with little interruption.[114] Moreover, contrary to lines constructed under the 'old system' station buildings and staff headquarters were of the most rudimentary sort. According to Seymour, 'stations, in accord with policy, were of the simplest kind – a corrugated iron hut about twelve by eight feet for the station office and with a few small round bush huts for the staff'.[115] This was, indeed, a stark contrast to the impressive two-floored brick station buildings made from imported materials that Shelford & Son had designed for railways elsewhere in West Africa.[116]

The railway built under the local system was completed in 1911 and cost £3,915 per mile, which was almost 40 per cent below an estimate of £5,679 prepared in 1905 by Shelford & Son. Despite the lowered engineering standards the quality of the railway was approved by inspectors sent to Nigeria by the Colonial Office.[117] In 1912 Lugard returned to Nigeria and future railway construction was placed in the hands of the public works department with Eaglesome as director of railway and works – a position for which in Lugard's opinion 'no man, however experienced in England, would be of any use unless he had at least five years of colonial experience'.[118] When the colonies of Southern and Northern Nigeria amalgamated in 1913 with Lugard as its first governor-general all future railway construction in the colony was placed in the hands of the public works department. The system of local construction was soon thereafter copied and adopted in the other and smaller British colonies in West Africa.[119] Hence, after 1914 in this region the services of Westminster consulting engineers were only employed in connection with telegraphy, electricity and waterworks – engineering fields where local expertise was still considered inadequate.

The developments in West Africa expose the conditions under which the use of Westminster consultants in Crown Colonies was likely to come under strain. Unsurprisingly, it caused friction if consulting engineers provided poor services as Shelford & Son clearly did. Moreover, controversies were likely to occur if the consultants not only prepared technical designs and conducted supervisions but also got involved with contractors' obligations such as the organization of the works and with labour supplies – as did Shelford, who even boasted before the Liverpool Chamber of Commerce that he was the biggest employer in West Africa.[120] Colonial administrators did not welcome interference in these issues as large infrastructural projects placed severe burdens on the limited resources of Crown Colonies. Indeed, leaving construction processes in the hands of consultants in London was likely to be considered an intolerable hindrance on administrators' ability to govern the colonies. As Girouard wrote to Lugard – who certainly shared the point of view – it made little sense that 'our railways are to be run by a man sitting 5,000 miles away' and it is not a personal question with either Shelford or Dunstan [a London-based engineer in charge of mineral surveys in the protectorate]; it is a matter of common sense and business, and a situation which would not be tolerated for a day by a self-governing colony.[121]

Conclusions

The expansion of the operations of the Westminster consulting engineers into wider geographical regions of the world mirrored the timeline of infrastructural construction across Britain's imperial system. From the mid-1860s and over the next two decades the London-based consulting engineers were mostly hired to projects in Canada, Australia, South America, India, the South African colonies and Egypt. Only from around 1890 was tropical Africa also drawn into the orbit where their services were requested. Indeed, in Africa engineers were more closely involved with the forces of imperial expansion than anywhere else. Often engineers were service providers for other and more powerful imperialist groups. Yet, consulting engineers were agents to whom large infrastructural projects were a main concern, the source of their income, their road to prestige and high positions. They were politically and financially interested agents and consequently did not just wait for business to come their way. Rather, from their position in Westminster they entered into alliances that exerted profound influence on the course of British imperialism in Africa. Baker's intermediate role in the financing of the Aswan Dam and Douglas Fox & Partners' connections with the BSAC are illustrative examples of this.

By the time the services of engineers were in demand in Africa, Westminster was well established as the nucleus of the British civil engineering profession and in particular as the home of the consultants who constituted the Great George Street Clique. These engineers were in a position to dominate and divide the burgeoning market for engineering projects in Africa. This they did. Those who were let in on the projects shared a number of characteristics and belonged to a circle of engineers who enjoyed the benefits of participation in networks based on strong ties and trust. This was in many ways their most valuable asset. Imperial projects constituted a vital force in shaping the development of these networks. Roles that had been defined and alliances that had been forged for projects on the domestic scene were adapted to a situation in which the services of the consulting engineers were required in all parts of the world where Britain's imperial influences were felt. It is a strong testament to the strength of the networks that they were capable of being maintained and adapted for projects throughout the imperial system. The endurance of the networks, moreover, owed much to the patronage structures and the bonds of trust that underpinned them. These factors were to the benefit of the insiders but naturally to the detriment of the prospects of engineers outside the Great George Street Clique. It was difficult for outsiders to change this situation. Indeed, as illustrated in the case of the Nigerian railways, it took a sustained effort from an influential opposition to oust the consultants from colonial projects even in cases where the services provided were clearly substandard.

The analysis of the enduring influence of the networks of the leading Westminster engineers has important implications for our understanding of developments in the late-Victorian engineering profession. There is little to indicate that increased professionalization in the engineering profession in the shape of growing specialization, more theory-based practices and the introduction of university-based engineering education diminished the significance of network structures based on personal connections and in many cases kinship. On the contrary these network structures often became increasingly important as the businesses of engineers expanded. Many of the advantages gained from network participation – reduction of monitoring costs, exchange of information over large distances, feelings of solidarity, conveyance of status and trust, continuation of long-term relationships – were even more vital as the businesses increasingly operated on an imperial scale. Far from being archaic leftovers from a world that was gradually giving way to professional society and meritocracy, the network and patronage structures proved a recipe for future success for the established elite segment of the engineering profession. Thus, a linear story of traditional structures based on kinship, patronage and personal connections gradually disappearing under the pressure of professionalization, meritocracy and rising bureaucracy cannot account for this development.

4 EMPIRE IN THE INSTITUTION OF CIVIL ENGINEERS

[O]urs is an Imperial Institute not a local society.
Benjamin Baker (1895)[1]

This chapter analyses connections between the British engineering profession and Britain's empire through a study of the ICE in London. During this period the institution was reformed to encompass a membership that increasingly consisted of engineers spread across the British colonies. The analysis demonstrates that imperial factors deeply influenced developments in the Westminster-based institution and thus directly impacted in the centre of the British engineering profession.

The line of argument pursued is, however, more complicated than this. The architects of the reforms in the ICE were the Westminster-based consulting engineers whose networks and imperial platform were analysed in the preceding chapter. This chapter argues that the way in which 'the empire struck back' in the ICE was shaped profoundly by the agenda of the consulting engineers of the Great George Street Clique. One side of this agenda was to make the ICE the imperial centre of the British engineering profession by strengthening ties with the growing communities of engineers in colonial diasporas. When pursuing this agenda a main concern was, however, the retention in Westminster of the ability to control how, and on what terms, these ties were to be forged. Metropolitan control of the ICE was a key issue for the Westminster consulting engineers.

By exploring the dynamics of this agenda, insights are gained into the ways that colonial regions and the engineers' centre in Westminster interacted reciprocally. Specifically, these dynamics are identified and analysed in relation to three topics: firstly, a reform in the organization of the ICE that was introduced in 1896 and for the first time gave colonial-based members a place in the council – the governing body of the ICE; secondly, the establishment in southern Africa of local advisory committees directed from the ICE headquarters in Westminster; thirdly, the publication activities and biennial engineering conferences that the ICE organized and housed in this period and which drew engineers from all over the empire to Great George Street. The chapter opens by outlining why the ICE was of crucial importance in the British engineering profession in this period and by teasing out the imperial dimension in the demography of the membership.

The Civils and the Empire

By 1875 the ICE was well established as the leading engineering institution in Britain and the model that other engineering institutions sought to follow.[2] The decades that followed were marked by unprecedented expansion of the institution. We have already seen that 'the civils' – as the members of the ICE were called – had established and expanded their headquarters in Westminster in order to cater for a membership that grew rapidly. In 1880 the institution counted 3,000 in their ranks, and by 1914 this figure had risen to over 9,000.[3]

From 1878 there were three main membership categories in the ICE. The category of students consisted of persons between 18 and 26 under training to become engineers and who were or had been pupils of a corporate member of the institution. The second and largest group was designated associate members and consisted of engineers aged more than 25 with a regular education as a civil engineer and who had been engaged in civil engineering work for at least five years. The third main class was designated members and consisted of engineers over the age of 30 who for more than five years had been employed as resident engineer or equivalent – or who alternatively had practised independently, also for a period of five years.[4] Above the layer of members was the select group of members of council. The council was the governing body responsible for the institution's finances, its daily operations and for expressing the official opinions of Great George Street. The council was elected annually by associate members and members from a ballot list prepared by the sitting council. The council also nominated candidates among council members for the positions of vice-presidents and president.[5]

Throughout this period it remained the case that practical experience and personal connections were the paramount criteria for advancement in the ICE. In this regard things were, however, not totally unambiguous. The reforms in 1896 discussed below established a requirement that candidates for associate membership should show proof of adequate theoretical engineering knowledge either in the shape of a degree in engineering from an educational institution approved by the ICE or, alternatively, by passing an examination held at the ICE headquarters in Great George Street.[6] The scope of this change towards criteria based on formal education should, however, not be overestimated. The council reserved the right to leave this requirement out of account in individual cases and in all cases election to associate membership required that six full members vouched for the qualifications of the applicant from personal knowledge. The transfer from associate member to the class of member involved no formal tests but required that ten full members vouched for the qualification of the candidate.[7]

For civil engineers membership of the ICE mattered. One important aspect was the privilege of a title. According to grade of membership engineers were allowed to use the titles of 'Stud.Inst.C.E.', 'Assoc.M.Inst.C.E.', 'M.Inst.C.E.', 'Member of Council Inst.C.E.', 'President Inst.C.E.'. When civil engineers in this period signed their technical reports and public writings it was standard to add the membership status immediately after the person's name. To a profession with little established traditions of formal education and few formalized procedures for career advancement the ICE membership grades were seminal designators of merit and authority.

To 'the civils' the ICE served a number of other important functions. The institution was recognized as a learned society for engineering science. Bernard H. Becket's *Scientific London* from 1875, for example, contained a full chapter on the ICE – and in fact devoted more pages to it than were allotted to the Royal Society and the Royal Institution combined.[8] In this regard the ICE's sizable library and lecture hall constituted the central venues. In the lecture hall the civils presented papers on topics related to the science and practice of engineering. Presentations were followed by a discussion among the members present. Papers and discussions were afterwards published in the *Proceedings of the Institution of Civil Engineers* – the institution's journal, which in this period was issued in four substantial volumes annually. A growing number of awards, medals and prizes were given for papers considered particularly eminent and this offered engineers additional possibilities for distinguishing themselves among peers. The first and most prestigious of these awards was the Telford Premium, established in 1835 and named after the first president of the institution. Other prizes and awards were successively established: The Watt Medal (1858), the Stephenson Medal (1881), the James Forrest Medal (1897) – the latter named after the secretary of the institution in 1857–97. In 1934 the Benjamin Baker Medal was added to what by then was a long list.[9]

In addition to its role as a learned society, the ICE also functioned as a London club for professional engineers; a venue for informal socializing and gentlemanly conversation. In line with codes of gentlemanly conduct, no political discussions were allowed and questions of wages and conditions of work were also regarded as taboo. Nobody *worked* in the institution but this did not prevent the home of the civils from functioning as what Clifton & Porter have called 'an informal clearinghouse' for judging the potential and trustworthiness of fellow engineers.[10] Indeed, socializing functions were of seminal importance in tying the growing community of civil engineers together. Besides regular weekly meetings held during the session running from November to April, the institution organized a host of recurring events, including the annual dinner and the annual conversazione. It was during the latter that the president, in line with established traditions, gave his inaugural address. The conferences of the institution, analysed in closer detail shortly, were another way in which the ICE was utilized in efforts to bind together communities of engineers. Marsden and Smith are thus right to assert that throughout the nineteenth century 'the identity of the civil engineer remained closely linked to membership of the civils'.[11] Not only did the ICE foster and strengthen a sense of community among Britain's civil engineers, it also established sophisticated formal and informal hierarchies within this community.

While the importance of the ICE has been noted and discussed by historians, the imperial dimensions have not been adequately teased out in the existing literature.[12] The first of these imperial dimensions comes out in the analysis presented in the preceding chapters of this book: the governing bodies and prestigious chairs of the ICE, from the council to the presidency, were dominated by consulting engineers (Chapter 2, Table 2.3) who from their offices in central Westminster were engaged in imperial projects across the globe (Chapter 3, Table 3.1). The ICE was the home ground of the Westminster consultants; the most important formal platform for the informal networks of the imperial consulting engineers.

A second imperial dimension in the ICE consisted in the fact that a growing number of members were based in the colonial world. Indeed, a substantial portion of the engineers who flocked to the British colonies or to regions of the world where Britain's imperial influence was manifest were members of the institution in Westminster. The ICE had colonial members almost from its founding. As early as the 1820s, a harbour engineer from Canada, Thomas Burnett, was listed as the first member based overseas.[13] Such members were referred to as corresponding members or non-resident members. Initially, this category included all members living more than 10 miles from London General Post Office but from the last quarter of the century it increasingly referred to members based in the colonies.[14] For these engineers the ICE constituted an important link to metropolitan influence. The group kept growing throughout this period in tune with Britain's imperial expansion. By 1875 a significant minority of members of the ICE were in colonial diasporas. In connection with the 1896 reform, discussed in detail below, a survey was made of the geographical distribution of ICE members in the British Isles and across the empire. Of the 4,960 members (based in Britain or colonies) in all classes (excluding students) 1,272 (26 per cent) were based in London, 2,141 (43 per cent) in the provinces, 375 (7 per cent) in Scotland/Ireland and 1,122 (24 per cent) resided in the British colonies (excluding Ireland). The colonial segment of 24 per cent was distributed with 10 per cent in

India, 10 per cent in Australasia, 1.5 per cent in Canada and 2.5 per cent in the category 'all other colonies'.[15]

A survey of the total ICE membership conducted for this book for the years 1885, 1902 and 1912 sheds further light on the nature and development of the colonial diaspora among ICE members (Table 4.1). From this table a number of points should be emphasized. Firstly, the number of ICE members in the colonies grew substantially during these decades and even marginally outpaced the very steep growth in the overall ICE membership that also occurred in the period. Thus, in 1885 the total number of ICE members (excluding students) in British colonies was 721 (or 18 per cent of the total ICE membership) while by 1912 it had more than doubled to a total of 1,651 (or 21.5 per cent of the total ICE membership). Secondly, the largest group of ICE members in colonial diasporas resided in the Indian subcontinent. This remained the case throughout the period though the percentage dropped from 8.7 per cent of the total ICE membership in 1885 to 6.3 per cent in 1912. Outside India the membership growth was more substantial. Geographically, the colonial segment in the ICE thus became notably more widely distributed with increases after 1885 in particular occurring in Crown Colonies, in Africa, Canada and in Australasia. Thirdly, it must be emphasized that a substantial number of ICE members were found outside the boundaries of the British empire; in continental Europe, in Asia, in the United States and in particular in South America where significant concentrations of ICE members existed in the states under 'informal' British imperial influence. A notable case in point, if unsurprising given Britain's extensive informal sway, was Argentina where the number of ICE members grew significantly in the last quarter of the century, from 14 in 1880 to 118 in 1890.[16] As will be evident it was, however, the groups of ICE members located within the formal boundaries of the British empire that influenced developments in the ICE during these decades. Fourthly, it may also be noted here that after 1914 the colonial membership continued to follow the pace of the general growth of the ICE membership; in 1928, in connection with the centenary celebrations of ICE's Royal Charter, it was calculated that 2,500 of 10,000 members of the institution resided in the colonies. The centres of colonial engineering diasporas also remained the same as prior to the Great War with the biggest concentrations of civils in India (621 members), Australasia (611 members), Africa (507 members) and Canada (139 members).[17]

Focusing exclusively on the African continent, figures are equally illuminating (Table 4.2). The total number of ICE members based in that continent grew from 68 in 1885 (1.7 per cent of the total ICE membership) to 267 in 1902 (4.2 per cent of the total ICE membership) and further rose to 379 in 1912 (5 per cent of the total ICE membership). In 1885, of the 68 Africa-based ICE corporate members and associate members more than half were based in Cape Colony.

Table 4.1: Members of the Institution of Civil Engineers Based in the Colonies in Selected Years[18]

Colony	1885		1902		1912	
	Members in all Classes[19]	Total ICE Membership (%)	Members in all Classes	Total ICE Membership (%)	Members in all Classes	Total ICE Membership (%)
India	349	8.7	411	6.4	483	6.3
Australia	125	3.1	309	4.8	296	3.9
New Zealand	54	1.3	78	1.2	76	1.0
Canada	56	1.4	73	1.1	180	2.4
Africa	68	1.7	267	4.2	379	5.0
Other Colonies	69	1.7	183	2.9	237	3.1
Total Colonial ICE Membership	721	17.9	1,321	20.6	1,651	21.6
Total ICE Membership	4,017	100	6,414	100	7,632	100

Roughly half of these engineers were registered with addresses in or near the diamond town of Kimberley while most others were employed with the Cape Colony railways or the colony's public works department (as was also the case with the 16 corporate members and associate members based in Natal).

In 1902 the number of ICE members resident in Africa had increased four-fold to 267. The rise had occurred in the established centres in Cape Colony and Natal but also in the Transvaal where the notable membership growth in the Boer Republic during the 1890 (from 2 in 1885 to 49 in 1902) clearly was related to the mining industry on the Rand – a case to be discussed in closer detail later in this chapter. Furthermore, after 1900 ICE members were also found in small numbers in tropical Africa, more specifically in West Africa, Rhodesia, Uganda and East Africa (where 6 out of the 7 were registered at the Uganda Railway main offices in Mombasa Island). These were the areas in tropical Africa in which extensive railway construction had commenced during the 1890s. The rather substantial increase of ICE members in Egypt was also related to a specific project; the reservoir works in progress at Aswan where more than half of the group of Egypt-based associate members were congregated. In 1912 the number of ICE members in Africa had risen further to 379 with increases both in the established centres in the dominion of South Africa, in Rhodesia and Egypt but also in the public works departments in the West African colonies and in the Anglo-Egyptian condominium of Sudan.

The imperial demographics in the ICE were thus somewhat complex: on the one hand there was a concentration of London-based but imperially-minded consulting engineers who dominated the council of the ICE; on the other hand there were substantial and growing expatriate communities of engineers in colonial diasporas. Against this background must be analysed the reform in 1896, the establishment of advisory committees in the colonies, and the publication activities and conferences organized by the ICE.

The Reform of 1896

As the group of 'colonial' members grew it became a contested issue whether they were given adequate influence in the ICE. The civils resident in the colonies represented all three main membership groups (members, associate members and students) but prior to 1896 none of the colonial expatriate communities was represented on the council. They were thus barred from direct influence in the governing body of the institution. Dissatisfied with this situation members based in India in 1883 requested that the council should alter the by-laws of the institution in order to enable them to elect from India one member of council and to introduce the opportunity of voting via the postal services. The request from India pointed out that the present procedure of electing the council at the

Table 4.2: Members of the Institution of Civil Engineers Based in Africa in Selected Years[20]

Colony/State	1885			1902			1912		
	Members[21]	Associate Members and Associates	Members in all Classes	Members	Associate Members and Associates	Members in all Classes	Members	Associate Members and Associates	Members in all Classes
Cape Colony	11	25	36	28	78	106	20	52	72
Egypt	6	7	13	6	28	34	19	67	86
Natal	3	13	16	11	27	38	6	28	34
Transvaal	1	1	2	10	39	49	27	70	97
Orange Free State/ River Colony		1	1		4	4	1	5	6
Rhodesia				1	17	18	2	23	25
Nyasaland/British Central Africa					1	1	1	2	3
Uganda					2	2	1	3	4
Gold Coast				1	5	6	2	5	7
Northern Nigeria							2	1	3
Southern Nigeria							6	7	13
Sierra Leone					2	2		1	1
Sudan				4	3	7	1	15	16
British East Africa							1	11	12
Total	21	47	68	61	206	267	89	290	379

annual general meeting in London effectively prevented members residing far from London from exerting their right to vote, and that this unfortunate situation could easily be altered by allowing postal voting. At first nothing came of this. The council considered it necessary that members elected should be able to attend the regular meetings in Great George Street in order for the council to fulfil its governing responsibility effectively. It was therefore considered prohibitive that overseas members under normal circumstances would not be able to be present at the council meetings. The council furthermore stated that it would not favour a particular expatriate community over others. On these grounds the request from the civils in India was refused.[22]

The issue, however, kept resurfacing as the membership – and in particular the large colonial minority of it – increased. Eventually, the ICE responded by adjusting the organizational setup of the council and the electoral procedures for it. This was done in 1896 under the presidency of Benjamin Baker, who was the architect behind the reform. For a special council meeting in December 1895, he prepared two memoranda titled 'Constitution and Election of the Council' which laid out what changes he considered necessary and how these should be carried out.[23]

In the memoranda Baker pointed out the discrepancies in the existing constitution of the council by providing a list which showed 'the distribution of the members and associate members throughout the Empire'. The numbers revealed that engineers based in the empire constituted almost 25 per cent of the total membership without them having any representation on the council. This discrepancy was problematic in the eyes of Baker and he argued that the solution was 'to enable the council to prepare a balloting list which shall be representational of all parts of the empire and of all branches of engineering'. Baker thus opened the way for colonial representation on the governing body of the ICE, but in his memorandum he also insisted that the standard of council members 'should not be lowered' and that it therefore was imperative to leave 'the preparation of the balloting list entirely in the hands of the council as it is at present'.[24] At the special council meeting the president's proposal was moved by consulting engineer Thomas Hawksley and seconded by William Preece and Guildford Molesworth. It was carried unanimously by the council.[25]

Subsequently, legal advice showed that the changes to the organization were of a nature and extent that required the obtainment of a supplementary charter. It was only the second time a supplement to the original Royal Charter of 1828 was obtained from the British Crown and it was thus partly done in response to the growth of the number of members stationed in the colonies.[26] The changes to the charter in 1896 furthermore made it possible for members prevented from physically attending the annual general meetings in London to vote for council representatives by postal ballot or by proxy.[27]

Prior to 1896 the council had consisted only of engineers practising from Britain – and indeed almost exclusively from London. After the reform, however, for the first time engineers residing in the overseas dominions and colonies were placed on the balloting list for election to the council. The overseas members were nominated for election in two successive years after which a new name was placed on the balloting list. The ultimate power to suggest and select names to be placed on the balloting list remained – in accordance with Baker's insistence – in the hands of the council. The new procedures introduced were thus by and large equivalent to the changes opted for in 1883 by members resident in India – with the significant addition that members from other areas of the empire were also made eligible for the council. In the years after the reform, the council placed on the balloting lists colonial representatives based in the areas of the empire where the major groups of colonial ICE members resided. The first year Henry Deane from the Railway Department of Sydney and John Kennedy from Montreal were elected to the council.[28] In 1898 Thomas Stewart from Cape Town and Alexander Izat from Gorakhpur in the North-Western Province of India were also selected for the ballot list and duly elected at the annual general meeting.[29] Thus in 1898 Canada, the Australian colonies, Cape Colony and India were represented on the council.[30]

Through this measure the ICE sought to adapt its organization to facilitate a membership that was geographically dispersed across the empire. In his presidential address in 1895 Baker also made plain the point that the ICE had taken a new departure to represent British engineers not only in Britain but also throughout the empire:

> I have little doubt that, had the wise men who drew up the original Charter of the Institution foreseen that the 156 members then to be legislated for would grow to the present 6730, they would have provided greater elasticity in the Charter, both as regards the constitution and the election of the Council. Provisions would probably have been made for the representation on the Council of our Indian Empire, the Dominion of Canada, the Australian Colonies, and some of the leading engineering centres in Great Britain and Ireland, for ours is an Imperial Institute not a local society.[31]

The minutes of the council meetings show that in the years after the 1896 reform, the council found 'suitable' candidates and placed them on the balloting list. The public statements that the council made in the *Proceedings* expressed satisfaction with this new departure, as in 1898 when the council announced that: 'The representation of the council in distant localities where large bodies of members reside, inaugurated two years ago, may, it is considered, be advantageously extended.'[32]

In influential corners, however, reservations existed concerning the election of colonial council members. In his presidential address before the institution in

1898, John Wolfe Barry commented on the issue. While he insisted that 'these [colonial] members of this body [the council] are of the utmost value to the whole body, being of much use for reference from time to time, and being trustworthy sources of information and guides in local matters' and that the system 'has been already, and will be in the future, a great success', Wolfe Barry also warned that:

> we must not forget that on those members who are resident in London or its neighbourhood, and can consequently be assembled frequently at the Council table, the main burden of Government and the ultimate decision on the important matters must of necessity devolve. Thus, the system of local representation should not, in my opinion, be carried too far.[33]

Such reservations also influenced the actual elections to council. The procedure of electing representatives from the larger communities of expatriate members did not become institutionalized fully and systematically in the following years. In 1902, for example, the council had problems finding one suitable candidate among engineers based in India in spite of the fact that there were 411 to choose from. After some deliberations, a motion put forth by Horace Bell (resident in Westminster) was resolved that in 1903 no representative from India would be elected.[34] The year before only South Africa was represented by members residing in that region whereas India and the Australian colonies were represented through respectively Frederic Robert Upcott, whose address was at the India Office in Westminster and Cecil West Darley from the New South Wales Government Office also with a registered address in Westminster. Canada had no presence on the council at that year's election.[35] Even after the changes to the statutes were made in 1896, what kind of representation the large colonial expatriate communities enjoyed still varied – and occasionally they had none at all. Furthermore, the council decided on a policy 'in which members representative of India and the colonies be not as a rule nominated for election more than two years in succession'.[36] This policy was rigorously enforced by the council and candidates were thus barred from serving on the council for long stretches – a limitation that impaired their ability to influence decisions. The two-year time limit did not exist for domestic members of council.

In spite of the inclusion of colonial representation on the council, the organizational backbone still consisted of engineers based in and around Westminster. The clear Westminster dominance did not change fundamentally until the end the period under study (as can be seen from Table 2.1 above). Westminster dominance in the council in fact ran deeper than what is indicated by these numbers. The power to put together the balloting list was kept firmly in the hands of the council and the colonial members that the council placed on the ballot list for election often had direct personal relations with the Great George Street Clique.

This is evident from an examination of the representatives from the expatriate communities in southern Africa. The first member elected to the council after the 1896 reform, Thomas Stewart, began his career as assistant in John Wolfe Barry's Great George Street office before he went to the Cape Colony and made his name as a leading hydraulic engineer.[37] Wolfe Barry was president of the institution in 1896–8. When Stewart retired from the council he was replaced by John Brown, who obtained his first appointment with the Cape Government Railways on the recommendation of the Westminster consulting engineer, John Coode under whose auspices Brown had worked as pupil and assistant.[38] Coode was president of the ICE in 1889–91. When Brown died he was replaced by Charles Metcalfe, the head of Douglas Fox & Partners' extensive operations in South Africa and Rhodesia. Douglas Fox was president of the ICE in 1900–1. The Westminster engineers clearly used their influence to place on the ballot list people they knew and trusted.

In the years that followed the council kept expressing satisfaction with the colonial representation system introduced with the 1896 reform. In 1910 the system was subject to evaluation in the council and the report it submitted concluded that:

> With respect to the representation on the council of the great dependencies of the Empire, the committee find no reason to suggest a departure from the view originally adopted by the council that the best interests of the Institution, and it is believed, local feeling, generally are in accord with the present practice of electing members resident in these localities upon the council from time to time.[39]

This system did not, however, work to the satisfaction of all, and in particular not to the contentment of 'local feeling'. A glint of dissatisfaction is revealed in September 1902 in connection with a social gathering of the civils residing in the Cape Colony. The total number of members of the institution of all classes based in the Cape Colony was 106 that year (Table 4.2). The gathering was held to celebrate John Brown on his election to the council. In the coverage of the celebrations, *Engineer*, in an editorial, used this occasion to urge the ICE in London to do more for its expatriate members:

> There are hundreds of engineers scattered all over the world who have done and are doing valuable work, the nature of which qualifies them for membership or associate membership provided of course, that they can pass the necessary examinations and test, but who are unacquainted with the methods of the Institution.[40]

The editorial argued that 'when a few years ago [the 1896 reform] the institution of Civil Engineers decided to elect to the council certain colonial members it took a very wise step in the interest of engineering and of itself', but the editorial insisted that the reforms had not been far-reaching enough. Above all it was important that the nomination of candidates for election to the council should be left to the local communities rather than being decided upon in London where often little was known of the local conditions. *Engineer* also urged the institution to appoint local secretaries who would form the link between Great George Street and the communities abroad. This was needed as 'members of council are appointed from the top of the engineering tree and have not time for secretary work'. The editorial concluded that:

> at the present day, the Institution as a great national professional organisation, should carry on its work on imperial lines, and do anything in its power to strengthen the relation between its members in far-off countries, and to bring them more closely in touch with the central body.[41]

Thus in the eyes of the editor of *Engineer* the ICE needed to work on imperial lines if the institution was to function as a catholic national organization for Britain's engineers.

The ambitious council did in fact share with the editor of *Engineer* the vision of forming the institutional framework for British engineers across the empire. The inclusion of colonial representation in connection with the reform of 1896 can be seen as a way of incorporating and strengthening connections with the growing expatriate communities throughout the British empire; an empire in which Great George Street saw itself as the engineering centre. The council was not, however, prepared to adopt a course which handed over much influence to the local expatriate communities. The institutional ties with civils in the colonies were to be forged in ways that gave the council the ability to set the terms of engagements. The power resided in the imperial centre rather than with colonial expatriate communities, and from the point of view of Great George Street it should remain there.

It is not surprising that change in the organizational set-up was only introduced slowly and that the Westminster dominance remained paramount throughout the period. This was not only or even primarily because of the institutional conservatism bound to thrive in a professional association where appointments to the higher offices depended on lifelong service and personal relations. Rather, at the time a strong case could be made that the basic organizational platform was successful. The membership curve rose steeply in this period and the number of colonial members even more steeply. Seen from Great George Street, the ICE had managed to create a set-up along the 'imperial lines' that, for instance, *Engineer* argued it ought to. In fact, the Westminster council members were in some respects the more ambitious party: they intended to house a strong metropolitan, imperial institution, not a conglomeration of colonial expatriate communities.

Advisory Committees

There was room for tensions as the council members sought to establish links with the expatriate communities and at the same time retain the links on their own terms. It is against this background that the conditional inclusion in the council of 'trusted' colonial representatives in the years after the 1896 reform should be analysed. It is also the context in which the relations between the council and a set of institutional bodies in the colonies known as 'advisory committees' must be seen – relations that this next section explores with a particular focus on the case of southern Africa. This analysis opens the way for deeper explorations into the dynamics at the colonial end of the imperial relationships that shaped developments in the ICE.

Advisory committees to the ICE were established in the colonies from the early 1890s. Importantly, they were set up not on the request of the council but on the initiative of the expatriate colonial communities themselves. The seminal function of these advisory committees was to advise the council on election and transfer of ICE membership candidates residing in the colonies. In line with the council's imperial agenda advisory committees received encouragement from London while at the same time their powers to make decisions on their own were kept on a tight leash.

The history of the advisory committees can be traced to 1880 when William Bennett, a member based in Esquimalt in British Columbia, suggested to the council that corresponding members in Australia, India and Canada should inform the council on the merits of candidates for the ICE membership who were based in these areas. The point of view was shared by John Thornhill Harrison, a member based in Westminster, who demanded that engineers applying for membership on services rendered outside Britain should be vouched for by members with actual knowledge of them.[42] While nothing initially came of this, the concerns raised in 1880 were similar to those that in the 1890s led to the establishment of the advisory committees: the perceived need within the expatriate communities to assist or influence the council in the assessment of 'colonial' candidates applying for membership of the institution. This issue was particularly pressing in Australia and in South Africa – the areas where the growth in the number of ICE members was particularly steep during the 1880s and 1890s. Indeed, the Australian colonies led the way when in 1890 a dispatch from Victoria informed the council that a committee would be established with the intention of assisting the council in London in relation to the election and transfer of members residing in the colony.[43] New South Wales (1892), South Australia (1895), and Western Australia (1896) soon followed suit.[44] Thus, throughout the 1890s, the council in London was with increasing frequency approached by small groups or individual members in the colonies who worried whether decisions regarding election and transfer of engineers residing in their respective regions were based on adequate knowledge of the candidates, and who therefore wished to assist the council in this regard.[45]

In late 1897, the council decided that the ICE should form an official position towards the advisory committees, which by then had proven to be more than a passing fad. The council therefore set up a committee on its own consisting of the president and vice-presidents, the task of which was 'to evaluate and consider the role of Advisory Committees'.[46] Upon the recommendation of this committee the council adopted a stand on the issue of advisory committees which was at best partly supportive. It resolved that 'Advisory Committees, whilst assisting the council with the information as to the qualification of local candidates, should not in any case bar or postpone direct application to the council of persons wishing to join the Institution'. Propositional papers from candidates should still be sent directly to the institution, which then 'would in necessary cases seek the private advice of the committees'. The committees were not to be public representative bodies but should instead stay in confidential and private correspondence with the council. It was furthermore resolved that advisory committees should consist of no more than six members. The chairman of the committee was always to be the member from the colony elected for council, and this chairman would have the right to appoint two committee members at his own discretion. Members residing in a region should elect only the three remaining advisory committee members. This astute set-up meant that the 'trusted' member elected for council would also dominate the advisory committee. Financially, the local advisory committees were also kept on a tight leash as the council resolved that the committees in the colonies would not be allowed to seek subscriptions from the local members but instead would have expenses covered directly by the council.[47]

Thus, advisory committees were established on the initiatives of colonial members and did not enjoy unconditional backing from the council in London. A closer analysis of the history of the advisory committees in southern Africa demonstrates that this issue was a source of dispute and discontent among colonial ICE members. The origins of an advisory committee in South Africa can be traced to the mid-1890s. The council in London received the first enquiry from this part of the world regarding the issue in 1893 when W. M. Grier of the public works department in Cape Colony contacted London with the familiar intention of establishing a committee for the evaluation of local candidates.[48] Grier died shortly after and nothing immediately came of this. However, already the following year an advisory committee was set up not in the Cape Colony but in the Transvaal.[49] The driving force behind this committee was Sydney H. Farrar, a British civil and mining engineer who had established a successful business in 1886 on the Witwatersrand gold reefs.[50] When the council was first approached by Farrar there were fewer ICE members residing in the Transvaal than there were in the Cape Colony. According to Farrar, the number of civils in Transvaal in 1894 amounted to a total of 5 corporate members and 18 associate members in Johannesburg and, in addition, 14 associate members in the Transvaal outside Johannesburg.[51] However, Farrar argued that the issue of a local advisory committee was much more urgent in the Transvaal than in the Cape Colony. The reason was that in the Cape Colony candidates were usually in government employment and could be vouched for by responsible heads of departments whereas the engineers in the Transvaal were employed in private business – attracted to the Transvaal Republic by the gold magnet of the Rand. Farrar and the engineers who promoted the Transvaal advisory committee were clearly concerned that gold diggers would gain the authoritative stamp as members of the ICE without possessing proper qualifications. According to Farrar this matter was so urgent that the Transvaal had to take immediate action 'to establish the Advisory Committee but [it] may later be joined with a Natal and a Cape Committee.'[52]

Elections for the advisory committee of Transvaal were duly held in 1894 with the civils residing in the Republic constituting the electorate.[53] Over the next two years Farrar, on behalf of a committee consisting of five men, reported back to London at intervals with the authoritative 'local' view on candidates applying for membership. In 1896, for example, Farrar reported that Mr William Morris Prout had applied for associate membership and that the advisory committee after examining the case considered his qualifications satisfactory.[54] Shortly thereafter Prout appeared in the ICE membership lists in the category associate member.[55] As this incidence indicates, advisory committees were not devoid of influence and there were advantages to be gained from being in a control of them.

Farrar, however, lost his privileged position in 1896 when the Transvaal advisory committee fell apart on account of developments in South Africa. Following a failed coup to overthrow the government of Transvaal – the infamous Jameson Raid – a substantial proportion of the British section of the mining community had left or was forced away from the Rand.[56] Farrar's brother, the mining magnate George H. Farrar, was one of the Uitlander ringleaders supporting the *coup d'état* from within the Transvaal. When it failed he only escaped a death sentence for high treason when his brother paid a fine of £25,000.[57] According to Sydney Farrar only two members of the advisory committee still resided in the region nine months after the failed coup.[58]

The shaky platform collapsed under Farrar's feet shortly after. In December 1896 a battle for the control of the advisory committee was fought in Transvaal with Farrar on the losing side. At a meeting held in Johannesburg a new advisory committee was constituted with Edward Baldwin John Knox – a consulting engineer and ICE member based in Johannesburg – as chairman.[59] Farrar complained to the council that this meeting had been publicly announced and that this was not in accord with the council's intention that advisory committees were not to be public organizations but rather advisory bodies that the council would consult privately when necessary.[60] Farrar protested in vain. At a speedily held election of the Transvaal advisory committee – with 5 members and 53 associate members in the Transvaal forming the electorate – Knox was formally constituted as new Chairman of the advisory committee.[61]

The archival records show very little trace of any activity of this newly constituted committee and, given its turbulent immediate history, it appears not to have enjoyed the full backing and confidence of the council in London. Conditions were exceptional on the Rand in the 1890s where the mineral bonanza placed engineers in the centre of political, economic and indeed armed conflict. Yet, the incident does bring out part of the reason why the council in London was reluctant to divest much of its power over electoral processes to the growing communities of engineers in the colonies. In a world of colonial engineering diasporas the task of maintaining and protecting the authority vested in the name and titles of the ICE was not taken lightly. Measures were needed to ensure that information from colonial centres and frontiers passed through hands that the council in Westminster considered worthy of trust. This issue of trust was all the more urgent given the large distances that ruled out direct and continuous monitoring.

The inactive advisory committee constituted in the Transvaal in late 1896 with Knox as chairman had ceased to function before the outbreak of the South African War. Entirely new and more ambitious beginnings were made in 1904 when an initiative in South Africa was taken for the formation of a South African advisory committee this time covering Cape Colony, Natal, the Transvaal, the Orange Free State and Rhodesia as one territory.[62] Ballot lists were prepared and sent out to members in the region and an advisory committee was constituted in August 1904. Charles Metcalfe, who at the time was also representative for South Africa on the London council, was elected chairman of the committee in accordance with the election procedures resolved by the council in 1897.[63]

Over the following years elections were held and the South African advisory committee was at intervals privately consulted by the council in relation to the assessment of candidates. Several incidents, however, reveal that collaboration with London was anything but smooth running. In 1907 the South African committee contacted the council requesting to take charge of organizing examinations in the region for candidates applying for ICE membership. This was outright rejected by the council who would go only as far as allowing the chairman of the advisory committee to be present at the examinations.[64] The South African advisory committee was again overruled in 1911 when the committee was informed that its membership transfer of Mr A. P. Gibb of Johannesburg would not be accepted by the council as the committee's actions in the matter had been in disaccord with an unspecified section of the bylaws of the ICE.[65]

Frustrations climaxed in 1914 when A. M. Tippett resigned his chairmanship of the advisory committee.[66] Tippett was chief engineer to the South African Railways in Johannesburg and had been a member of the advisory committee since 1912. In his resignation he stated that the committee was of no use as its members hardly ever got together on account of the vast territories it was meant to cover. Furthermore, Tippett considered that 'the duties that the institution requires the committee to fulfil, appertaining as to the suitability of candidates, are of such trivial nature that they no longer appear to me to call for a committee'.[67] The reply from London expressed regret that Tippett had chosen to resign and insisted that:

> The primary object of the Council's advisory committees throughout the world is to assist them with regards to the qualifications of candidates for election into the institution and this [is] a matter of extreme importance to its interests and status.[68]

Alert to the fact that the South African advisory committee was again falling apart, the council persuaded Thomas Stewart – the veteran who had represented South Africa on the council in 1898 – to fill the position as chairman. Stewart, who was sick with malarial fever, reluctantly accepted though he expressed agreement with Tippett that a committee of such limited powers was not needed and that information instead could be gained directly from trustworthy members.[69] The dying South African advisory committee finally dissolved in 1915 after the outbreak of the Great War. It was not resurrected until 1927.[70]

Throughout the period, the council's support of the advisory committees was tepid. The committees remained purely advisory organs to the council with functions reduced to private communication on the assessment of potential candidates for membership. Moreover, advisory committees did not feature prominently within the ICE and only appeared in the published by-laws and membership lists after 1912.[71] This also explains why they have escaped scholarly attention – even in Garth Watson's detailed history of the ICE.[72] Committees were left with little room to manoeuvre, remained secluded and had limited authority. The election procedures established from London meant that the member of council for the colony was automatically elected to the chairmanship of the local advisory committee and retained ultimate control of it. From the point of view of the council's meeting table in Great George Street this measure was considered indispensable in order to ensure that information from the colonies passed through trustworthy hands from 'the top of the engineering tree'.

The analysis of the Transvaal and South African advisory committees strongly suggests that engineers in the colonial diasporas were not satisfied with the power and authority that Great George Street vested in them. However, as in the case of the 1896 reform, the unwillingness to hand over authority to the colonial expatriate communities of engineers did not result from lack of ambition within the London institution. On the contrary, it was the ambition to remain the engineering centre of the British empire that made the council members reluctant to give much power to members in the colonies.

Publications and Conferences

The council in Great George Street was reluctant to hand over direct power and authority to the expatriate groups of engineers. Instead, the governing body of the ICE endeavoured to maintain links with engineers in the colonial diasporas by other means. The strengthening of connections and communal bonds between Westminster engineers and colonial members was a prerequisite in fulfilling the roles of the ICE as these were envisioned in London – a vision referred to by Baker in his presidential address when he insisted that Great George Street was the home of 'an Imperial Institute not a local society'.

An important component in strengthening links and bonds consisted in systematic efforts to collect and disseminate printed material relating to engineering projects in the colonial world. Members of the institution were encouraged to send back writings on their work overseas to the ICE's library. In 1901 in Uganda, George Whitehouse for example received a letter from the ICE thanking him for dispatching to Great George Street reports from the Uganda Railway Committee.[73] The letter to Whitehouse was a standard printed template that must have been sent to other engineers abroad. It is difficult to establish how formalized such procedures were but the huge collection in the ICE's library of pamphlets, printed speeches, articles and reports produced by members on projects in all corners of the globe in this period tells its own story. By 1928, the library's collection of pamphlets alone numbered 16,700.[74] The existence of the vast collection is a strong indication that members considered the gathering of information important and that this process was encouraged from Westminster.

In terms of printed communication the *Proceedings of the Institution of Civil Engineers* was the cornerstone in the exchanges between the institution and engineers residing far from Great George Street. From the 1870s the *Proceedings* were expanded to make special provisions for this task; besides printing the papers and discussions from the lecture hall, the publications from then on also contained sections with incoming correspondence from engineers who had been unable to attend the meetings but instead had been encouraged by the ICE to comment on the presented papers.[75] In 1912, for example, Great George Street resident Frederic Shelford spoke in the lecture hall on 'Some Features of the West African Government Railways'.[76] Shelford's paper dealt with issues such as surveying, clearing and construction in the demanding environment of African rainforests. In the ensuing discussions engineers commented on the paper with references to their own experiences from India, South America, the Malay Peninsula and elsewhere. One discussant, Mr Humby – a member who remarked that he 'had about 20 years' experience of railway engineering in South Africa' – would

> particularly recommend the paper to the attention of younger members and students because the author's valuable descriptions of the various methods by which he had overcome the difficulties arising in the field were not to be found in elaborate textbooks.[77]

When Shelford's paper was published in the *Proceedings*, it was followed by 36 pages of incoming correspondence from civils residing in the colonies, who drew comparisons with railway construction in Australia, Canada and in other parts of Africa.[78] The printed *Proceedings* were distributed to educational institutions, technical libraries, professional societies all over the world, and – it appears – upon request to individual ICE members.[79] In this way the ICE provided a platform for the exchange of experiences gained in diverse geographical settings.

This exchange was not only important for sharing technical advice and knowledge. In the apt words of the historian of science J. A. Johnson, publication for colleagues constitutes 'a fundamental social act'. By establishing regulated forums for communication, it can serve a seminal role in the alignment of practices and is thereby instrumental in shaping the professional identity of a communicative group.[80] Executing the task of compiling, publishing and disseminating written materials in formalized ways was part of the platform of the ICE for binding together the community of members beyond the lecture hall and across the world. A further indication that this task was high on the agenda in Great George Street can be identified in 1906 when an annual Overseas Premium Medal was instituted and awarded to corporate members for the best paper presented on works carried out overseas.[81]

Community-building efforts were also a notable feature of the conferences held in the ICE. In this period, general engineering conferences were organized in 1897, 1899, 1903 and 1907 and drew engineers to London from all corners of the empire. These conferences have hitherto escaped historians' attention. However, they constituted a seminal way in which the ambitious institution sought to establish its position as the engineering centre of the British empire. The idea of engineering conferences held under the auspices of the ICE originated in the council. In the spring of 1896 the newly constituted president, John Wolfe Barry, successfully put forward a proposal for 'an annual gathering or conference of the members of the institution to be held in London'.[82] All conferences were organized by the council through a committee consisting of the president and vice-presidents in office during the years of conference.

From the outset the gatherings were regarded as a huge success. At the first assembly in 1897 more than 800 members participated and the number of attendants remained stable for the next conference.[83] In conference years the annual conversazione of the ICE was held in connection with the event and here around 3,000 participated – mainly ICE members but also distinguished men from the worlds of politics, finance, industry and science. Apart from dinner and musical entertainment, the conversaziones were also used for displays and exhibitions of scientific and technical novelties and products. The new tradition of the conference conversazione climaxed in 1907 when the event was moved to the grand Royal Albert Hall.[84]

The basic format of the conferences was the same throughout the years. The gatherings opened with a plenary session during which the presiding president of the ICE addressed the large audience of members. After the inaugural session the conference split into sections according to topics that had been selected by the council's conference committee in collaboration with a select group of members. These topics remained the same during all four conferences. During the first conference in 1897 the section on 'Harbours, Docks and Canals' was chaired by Harrison Hayter, 'Railways' by Benjamin Baker, 'Machinery and the Transmission of Power' by Frederic Bramwell, 'Mining and Metallurgy' by Thomas Forster Brown, 'Shipbuilding' by William White, 'Waterworks, Sewerage and Gasworks' by James Mansergh and 'Application of Electricity' by William H. Preece. Apart from Forster Brown all chairmen were Westminster-based consultants.[85] Each topic for discussion was introduced with brief notes produced by select members in a manner 'less formal' and with 'a more conversational style of discussion admitted than was advisable in the ordinary meetings'.[86] The idea was to facilitate discussion and to allow informal interaction between professionals who rarely got together. Consequently, the notes were kept short with a limit of 1,000 words formally introduced in 1907.[87] The informal and social elements of the conferences were given strong priority. The discussion sessions lasted only three and a half hours daily from 10 a.m. to 1.30 p.m. The afternoons were reserved for excursions to various engineering works. During the first conference, the excursions were to sites in and around London such as the projects for the purification of the Thames carried out by John Aird & Son.[88] In 1903 trips were organized as far as to Crompton & Co.'s works at Chelmsford and to the project carried out to expand the harbour at Dover.[89] These excursions can best be seen as a professional variation of what Marsden and Smith have labelled 'technological tourism'.[90]

The professional and the social elements were most intimately intertwined in connection with the conversaziones. On these occasions dinners, speeches and performances by the string band of the Royal Engineers mixed with exhibits of new technologies and instruments.[91] In between wining and dining during the conference in 1897, the members could stroll by the stand of King's Norten Metal Company of Birmingham which had on display a number of solid drawn cartridge cases of various sizes. Or they may have chosen to inspect the complete model of a Robey Winding Plant with a two-cylinder, semi-portable engine and boiler that was presented by John Richardson. A third opportunity afforded was to become acquainted with the railway surveying instruments that J. C. Ford had on display in the library. These were only a few of the many exhibited objects that according to *Engineer* 'would be of much interest to any member' of the vocation.[92]

By including such displays the ICE conferences embraced the 'Cultures of Exhibition' that have long held the interests of scholars concerned with the relations between British imperialism and British culture.[93] It has been noted that in the imperial exhibitions 'trade, technology, and progress were central motifs' that linked the practical and material concerns with 'goals of peace, imperial achievement, and unity'.[94] At exhibitions 'machines-in-motion' featured prominently and served to underpin visions of British ingenuity and superiority.[95] This idea was not absent in the ICE whose members generally were alert to any 'civilizing' potential of technology. The engineering conferences were a meeting point for professionals to whom the utilization of overseas resources was a central concern. Thus, although the engineering conferences lacked the size and the exoticism of the imperial exhibitions held in London and elsewhere during the era of 'New Imperialism' they were of no little importance. They served as a professional and social platform for people whose business and livelihood was closely linked to the empire. It gave people with skills and experiences developed in various places in the world a rare chance to exchange points of views in relation to issues such as, for example, 'Location and cost of working Pioneer Railways' or 'Reciprocating Engines for Ocean-Going steamers'.[96] These were the machines-in-motion with which colonial empires were built.

In the engineering press the coverage of the conferences was generally positive. The *Engineering Magazine* in its editorial columns spoke highly of the 1899 conference and concluded that it 'will be remembered as a most important professional and social gathering'.[97] The big weeklies, *Engineer* and *Engineering*, were the most eager conveyers of news from the proceedings in Great George Street. Well in advance the full conference programmes were printed in both journals; abstracts and extracts from the notes were carried, with *Engineering* publishing more of these than *Engineer*. The president's opening address also was reviewed and extensive reports covered the exhibitions and displays in connection with the conversazione. *Engineer* made its own departure by running a series of articles on the engineering sites visited by the conference participants.

Leading articles further endorsed and praised the institution's initiative and confirmed the ICE's status as 'the mother institution of the profession'. In 1903 *Engineering* claimed that 'these conferences emphasize the great catholicity of the parent institution of our profession' and the editorial went on to reflect on the national importance of the gatherings:

> These bi-annual stocktaking's [*sic*] in the departments of engineering knowledge cannot fail to have a stimulating effect on industry. They show the direction that events are taking, and direct the minds of engineers into useful channels in which their labours are likely to be remunerative. They also serve as warnings to those who are inclined to stand on ancient ways, and who do not sufficiently realise the necessity of action and progress. As far as such persons are concerned it is not a matter of great importance whether they be aroused or not; but from a national point of view there is the most pressing necessity that every engineer should be fully cognisant of the latest improvements and the most recent victories of science. It is better to learn such lessons from each other in council than from the intrusion of foreign manufactures into our midst.[98]

Importantly, the conferences were framed as both national and imperial events. When the idea of a conference first caught on in the council, it was seen primarily as a means of strengthening ties with engineers outside London but still residing in the British Isles. In 1896 this group made up nearly half of the membership at the time, while the colonial segment, as discussed above, constituted roughly 25 per cent.[99] When the council at the annual general meeting in 1896 made public the intention to host a conference the announcement stated that the council

> proposes to hold a general conference of the Institution in London on the 25th, 26th and 27th May next. It is hoped that this opportunity for men practicing in various branches of engineering in all parts of the kingdom, to meet in the Metropolis for the interchange of professional views, may satisfy a need that has been felt in an increasing degree with the expansion of the Institution.[100]

This 'domestic' line of argument was also taken in the opening address to the conference, where Wolfe Barry declared that the object of the conference was 'to unite more closely' London with 'the Great Towns of the Kingdom'.[101] The press also placed more emphasis on the national than on the imperial side of the conference, though *The Times* did briefly note 'the presence in London of a large number of foreign and colonial members'.[102] *Engineer*, however, expressed the more common view when an editorial on the conference declared:

> One consequence of the fact that large numbers of members and associates reside permanently in the country has been a complaint, felt perhaps rather than expressed, that the Institution was not quite in touch with a large proportion of its members. Men living in provincial towns yet holding important positions only rarely had the chance of attending the meetings or entering the home of the Institution in Great George

Street. Essentially a London Institution, the country members were in a measure shut out from its advantages; and as the Institution never left the metropolis, if a change was made for the better it was apparent that the members must come to the Institution. It was, in short, the old case of Mahomet and the Mountain. The first effort in this direction has taken the form of a Conference, and how popular this promises to be may be gathered from the enormous number of applications which have been sent in for tickets.[103]

The imperial element was in the background of the agenda for the conference in 1897. However, it took centre stage at the next conference held two years later. Here the organizers placed emphasis on the wider and explicitly imperial reach of the conference. In his opening address in the lecture theatre of the ICE in May 1899 the president, William H. Preece, explained how and why this was the case. He stated that the aim of the conference was to 'facilitate intercommunication among our members by bringing us more into personal and social relations with one another'. This was also a central purpose of the ordinary weekly meetings in London, but

> these Conferences, in association with our annual conversazione and by combination with well-arranged excursions and visits to works, tempt many to come to Westminster who are rarely seen here, and facilitate the interchange of ideas between all the departments of our many-sided profession. They disseminate knowledge by exciting discussion, and extracting opinions from those who have acquired them from the only true master – experience – in every corner of the globe.[104]

Preece referred to civils working 'in every corner of the globe', but his opening address explicitly imbued the conference with an imperial – rather than a national or international – meaning and significance:

> We are gathered together from all parts of the Kingdom to hold our Second Metropolitan Engineering Conference; and we welcome here many, who, temporarily at home, pursue their ordinary work in distant parts of the Empire, or in foreign countries ... We embrace in our sphere of operation the whole world. We desire to make our home in Great George Street the Mecca of Engineering; and in furtherance of this idea, I am bold enough to suggest that, at the next Conference, the Council should take measures to secure the presence of some of our members, delegated specially to represent Engineering in our Empire beyond the seas. This Conference is not international, in the sense of that held at Chicago in 1893, or of that which I understand is contemplated for the year 1901 in Glasgow, in connection with the Exhibition to be held there; but it may well be Imperial. In whatever quarter of the globe we find ourselves, membership of this Institution has become a pass-word; and the invariable support accorded to the Institution by its members in India and the Colonies suggests the desirability of uniting them in a definite and formal manner with our proceedings on such occasions as this.[105]

Preece explicitly contrasted the conferences of the ICE with the international engineering conferences that were held in Chicago, Glasgow and elsewhere in the period by highlighting the imperial dimension. The international aspect was not absent in Great George Street but the event was above all imperial just like the bonds were between the civils that it was meant to build on and strengthen.

By hosting the well-attended conferences the ICE had found another way to cement its position as 'the Mecca of Engineering' in Britain and the empire both among the members at home and abroad as well as in the wider public. Indeed, Preece made his point to an even larger audience than the one present at the conference opening. In an article in *Feilden's Magazine* – the 'Militantly British' engineering review magazine – Preece repeated that the conference was a service to the 'many colonial members that are always to be found temporarily at home on holiday or business, and anxious to learn all they can of home experience and up-to-date facts'.[106] Preece's contribution to *Feilden's Magazine* was the opening article in the very first issue of the review and it presented the president with an attractive opportunity for restating all the good that had come out of council's initiative to organize and host engineering conferences.[107]

Conclusions

During this period roughly 20 per cent of the rapidly growing membership of the ICE was based in British colonies. This fact underscores that the diaspora of British engineering was an imperial affair. Buchanan argued along the same lines but based his notion of the diaspora on a survey of the select few engineers who made it into the *Dictionary of National Biography*.[108] The survey of the number and activities of the colonial members of the ICE, thus, places this point on a more solid empirical footing.

Moreover, the numbers have provided the platform for exploring the dynamics produced by and during this diaspora. While the Indian subcontinent continued to house the largest group of colonial ICE members at this time the membership curve rose more steeply elsewhere in the empire, notably in the African continent where the number increased more than five-fold during this period and by 1912 constituted 5 per cent of the total ICE membership. The colonial diaspora of the civils in several respects had a profound impact on the institution in Westminster as engineers in the colonies aimed to influence the ICE for it to take greater heed of their specific needs and requirements. They did so by pushing for the inclusion of representatives from the largest colonial expatriate communities into the council of the ICE and by setting up advisory committees in the regions they resided in. These initiatives can be viewed as tangible ways in which the empire 'struck back' in the ICE.

The imperial influences in the ICE were, however, at least as much metropolitan in origin and nature as they were colonial. The organizational set-up and outlook of the ICE was dominated by the Westminster consulting engineers, and it was their imperial vision that played first fiddle in the Great George Street home of the civils. Indeed, for the consulting engineers of the informal network of the Great George Street Clique the ICE was a powerful formal platform. High on the agenda of the consultants was the aim to make Westminster 'the Mecca of Engineering' (to quote William Preece's apt phrase) of the British empire – an empire to which their own business platform, professional status and identity was intimately tied. Pursuing this agenda meant retaining and strengthening connections with British engineers in colonial diasporas. When these connections were established the council members relied on the people with whom they had 'strong ties of friendship and obligation'[109] and in all cases the connections involved an uneven distribution of power and authority.

The growing importance of the empire in the ICE during these years owed more to the small but centrally placed group of Westminster consultants than it did to the large expatriate communities in the colonies. The metropolitan engineers settled the terms for the reform in 1896. They were also capable of dominating, and eventually sidelining, the colonial advisory committees, as was shown by the case of the committees established in southern Africa and later shut down again in spite of the fact that a growing number of members resided in the region. Instead, initiatives were launched from the council room in the shape of new publication procedures and conferences that aimed at confirming the imperial connections between Westminster and engineers in the colonial world. Such initiatives served both internal and external needs. Internally in the profession, the publication format aimed at providing a communication platform for a geographically dispersed membership while the conferences gave engineers who rarely got together an opportunity to compare notes, take stock and discuss issues of communal interest. Externally, the conferences provided engineers in the ICE with a chance to lobby politicians, to have their presence noted in the press and public and to display their contribution to the empire and thereby to have it gauged and confirmed.

Thus in this period the organizational set-up of the ICE – the oldest, largest and most distinguished accredited body for engineering in Britain – was geared to encompass a profession that both practically and ideologically became more closely tied to the empire. This was, moreover, important beyond the walls of the institution as the identity of British civil engineers remained closely linked to the membership of the ICE. Indeed, a seminal component in the imperial identity of British civil engineering originated from the institution in Great George Street.

5 EXPLORER-ENGINEERS AND GENTLEMEN IN THE PUBLIC EYE

A powerful air of kinship pervaded that dinner. Each and every guest was filled with the sense of a tough job well done. Their limbs ached slightly, as though they personally had shovelled earth and hauled sleepers into place.

Peter Høeg (1990)[1]

The effects that engineering in the empire had on the British engineering profession were particularly strongly felt in the public spheres. Imperial projects and connections gave engineers a range of new opportunities among the British public and brought about changes in public perceptions both of what it meant to be a British engineer in Africa and of the frontiers the engineers claimed to be taming. This chapter analyses the ways in which engineers made use of these opportunities and identifies the reasons they had for pursuing them. An issue of self-representation is involved here, and the way engineers fashioned themselves and their contribution to the British empire greatly differed according to their standing within the complex professional and social hierarchy that was in place in the profession by the last quarter of the nineteenth century.

The position of individual engineers within this socio-professional hierarchy profoundly influenced the motives as well as the means they had for engaging with the public. Some engineers presented themselves as 'explorer-engineers', a term introduced in the first section of the chapter to label a cultural persona that rose to prominence as familiar images of 'Smilesian' heroic engineers fused with the vibrant late-Victorian cultures of empire and exploration. Other engineers and notably consulting engineers approached the public as gentlemen experts on colonial development and policymaking. The Westminster-based consulting engineers were among the most active engineers in public spheres and this was a part of their function as intermediaries or bridgeheads between Britain and engineering projects in Africa. The chapter demonstrates that the privileged access consultant engineers had to communication channels enhanced their ability to exert influence on developments in both locations. Moreover, for independently practising gentlemen engineers, who according to professional codes of conduct were not allowed to advertise their services, public exposure was a great attraction and the chapter argues that the public strategies of the consultants should be analysed as part of their trust-building efforts.

The Explorer-Engineer

Historians of science and technology have shown that during the nineteenth century a number of cultural images of engineers had developed, making the 'heroic engineer' a well-established if contested icon in Victorian culture.[2] In particular, the inventors and engineers whose names were associated with the transformation of Britain into the leading manufacturing nation and 'workshop of the world' became the subjects of numerous portraits and eulogies, most prominently in Samuel Smiles's immensely popular books on the lives of eminent engineers published in the 1850s and 1860s.[3] These 'Smilesian' engineering heroes were portrayed as self-reliant, hard-working and individualistic men of humble background who, unspoiled by formal, theoretical education and book learning, had a developed a talent for devising practical solutions.[4] Challenging and overcoming physical barriers in the British Isles as well as the social barriers that stood in their way, the engineers of the Smilesian mould were accredited with having produced faster means of transport, cheaper goods, employment and better living conditions for their fellow countrymen. The heroic engineers triumphed through their stamina, high moral integrity, noble ideals and solid character, and were described as the embodiment of the immensely influential and popular Victorian philosophy of self-help.[5]

Traditionally, historians have argued that the popularity of the engineering heroes reached its zenith in the mid-Victorian period – the Great Exhibition in London in 1851 and the almost simultaneous deaths of Brunel, Locke and Stephenson in 1859–60 marking convenient demarcation lines – and from there faded in tune with what in a once-influential historiography was labelled 'the decline of the English industrial spirit'.[6] More recently, several historians have challenged this view and demonstrated that inventors and engineers were the subject of complex cultural interpretations and appropriations that do not fit dichotomic notions of ascent and decline.[7] Most forcefully, Christine Macleod has shown, in a study impressive in scope and detail, that engineers and inventors were celebrated far beyond the pages of Smiles and that they remained important, if ambiguous, cultural icons also after 1860. Macleod also demonstrates that towards the end of the century images of inventors and engineers toiling in their workshops or in the British countryside had trouble competing for public attention with other iconic figures, most notably explorers and soldiers. She notes, however, that 'Although engineers active after 1860 failed to capture the imagination of contemporary biographers or historians, they did enjoy some imperial limelight'.[8] Macleod's astute observation holds true. Indeed, engineers enjoyed much more imperial limelight than has hitherto been recognized. In particular, the late Victorian period saw the emergence of a distinct cultural persona combining the characteristics of the Smilesian engineers with those of the celebrated explorers.

It is well established that few groups in the second half of the nineteenth century could compete for public attention and admiration with the explorers.[9] Like engineers explorers were, however, contested and ambiguous figures in Victorian culture. Felix Driver, among others, has pointed out that in this period the cultural icon of 'the explorer' underwent constant appropriations and this process has continued throughout the twentieth century.[10] Several groups associated themselves with exploration and laid claims to follow in the footsteps of icons such as David Livingstone and the more controversial Henry Morton Stanley.[11] Field scientists, doctors, travel writers, tourists and others would often in subtle and ambiguous ways relate their activities to those of the explorers who had 'trotted the field' before them.

Engineers constituted another group laying claim to being the heirs of the explorers. Central to this view was the idea that engineers were representative of the next phase: where explorers had taken control of the frontier by mapping it, engineers were now bending it to commercial exploitation and political subjugation by establishing infrastructure. The fact that engineers from the third quarter of the nineteenth century were hard on the heels of explorers and mapmakers added much to the persuasiveness of this view; settling railway trajectories or surveying potential mining areas often meant travelling along routes and through regions where few (white) people had gone before. There were overlaps in the field activities of engineers and the famed explorers – and also accompanying ideological overlaps. Livingstone's credo of civilization, Christianity and commerce was conceived to be open to scientists, soldiers, merchants and indeed engineers who (as was argued in Chapter 1) could describe themselves as taking part in this mission by emphasizing a fourth C of civil engineering: in this view infrastructure would create political stability, strengthen the position of missionaries, encourage investment and commerce and thus promote 'progress' in Africa.

The practical and ideological overlaps became even clearer in the more violent phase of exploration which Driver convincingly has argued gradually became dominant. It is worth recalling that Stanley, the controversial 'geography militant' icon of this period, earned his nickname *Bula Matari* – the breaker of rocks – from blasting away the rocks blocking the road that he had been employed by King Leopold to construct as a forerunner of the railway connecting the Atlantic Coast with the navigable sections of the Congo River.[12] Stanley was an explorer who also took on the role of the road and rail builder. Many engineers made the same journey but from the opposite direction to Stanley. By 1875 the ideological as well as the practical distance between pathfinder and railway builder was shorter and more blurred than ever. These were ideal conditions for explorer-engineers to flourish.

One consequence of this was the creation in British culture of an identifiable image of the engineer in Africa, an image that was fed through a number of channels and by a number of agents: by observant editors and publishers of popular magazines and technical journals (such as those examined in Chapter 1) and by prolific authors, as for example Rudyard Kipling and the young Winston Churchill, who expressed a widespread sentiment when he claimed that:

> Civilisation must be armed with machinery if she is to subdue these wild regions [of East Africa] to her authority. Iron road, not jogging porters; tireless engines, not weary men; cheap power, not cheap labour; steam and skill; not sweat and fumbling: there lies the only way to tame the jungle – more jungles than one.[13]

Even more important for the dissemination of ideas of engineering in Africa were journalistic writers such as Frederick A. Talbot, whose handsomely illustrated and popular book series on *Railway Wonders of the World* and *The Railway Conquest of the World* in chapters on 'the romance of construction' and 'the adventurous life of the railway surveyor' pictured engineers as conquerors of the globe, annihilators of space and time, civilizers of 'savage natives' – as 'the advance guard of civilisation'.[14]

Little is yet known about the scale and changing faces of this under-researched phenomenon at the cultural intersection of engineering and imperialism in a period spanning at least fifty years. However, as a cultural icon the explorer-engineer was strong enough to constitute an essential cultural backcloth against which engineers could present themselves and their profession in the public sphere. Indeed, engineers in a number of ways actively and increasingly associated their work and public personae with the vibrant Victorian and Edwardian cultures of empire and exploration.

Autobiographies were one channel for doing this. The first example in this genre may have been John Hawkshaw's 1838 *Some Reminiscences of South America*, describing his experiences as a young mining engineer in Venezuela.[15] Hawkshaw's book predated the era of the 'new imperialism' but it was a specimen of a genre that remained popular for over a century and which often commented on the role and character of the engineer venturing into the imperial field. Charles Burge, in his 1909 autobiography on *The Adventures of a Civil Engineer*, provided a typical character sketch of the frontier-going British engineer:

> Here [Decca Province in India], I made my first acquaintance with the typical constructing engineer abroad, and after meeting him since in many other lands, I may say that he is *sui generic* among his craft. He differs in many respects from the type I had left at home. Sunburnt, bearded, with the pipe ever in his mouth, a daring rider, full of energy, exhaustless in resource when difficulties arise, hospitable to the last degree,

and full of queer anecdote, he has little tolerance for fussy namby-pambyism in his superiors or his comrades, and expects his men to work as hard as himself, with due allowance for the various disabilities of the races with which he has to deal.[16]

According to Burge this type of engineer was resourceful and accustomed to dealing with tropical perils and people of different races – *sui generic*, in particular through his stamina and independence when faced with these challenges.

Major engineering projects in the colonial world were collective efforts involving hundreds and occasionally thousands of people but the writings highlighted the contribution of the 'lone heroic engineer'. Scholars who have analysed the writings of explorers have demonstrated that the collective nature of exploration was usually downplayed and in particular that the often crucial contributions of indigenous people almost disappeared from the published narratives.[17] This was also the case in the writings of explorer-engineers; when they triumphed they usually did so in spite of their workforce rather than in collaboration with it. Indeed, generally the engineers' writings reflected the racial and cultural prejudice of a period in which indigenous peoples in Africa were viewed as highly inferior in particular with regard to their abilities to handle 'Western' technologies.[18] This aspect featured in a range of ways. Percy Girouard, the Royal Engineer, colonial administrator and prolific writer who first made his name as 'Kitchener's railway man' during the Anglo-Sudanese war and later replaced Lugard as high commissioner in Northern Nigeria, in his autobiography expressed very little patience with the Egyptian and Sudanese workforce he felt he had to make do with:

> Never had a motlier crew been brought together in the guise of a Railway Battalion ... If the best theodolite goes smash upon the ground, if the locomotive runs without a gauge glass and burns a fire box, if the precious belongings are lost and the white mans' Western ire is raised – 'Why oh why Excellency be perturbed? Ten thousand years ago it was ordained I should do these things this day' ... The Egyptian fellahin, his nearest prototype in fatalism is the Chinese, his modern prototype the Russian moujik.[19]

For Girouard there was a clear and direct link between the inability of the Egyptians to appreciate even the simplest 'Western' tools and the invariable superstition and fatalism they suffered from. Frederic Shelford, to whom we shall return shortly, expressed a similar view but also warned 'the pioneer engineer' not to

> imagine a native carrier is a mere beast of burden, without soul or feeling. He is after all, a human being, and there is a distinct difference between the services rendered by a well-treated native and an ill-treated one. It is easy, by gentle but firm dealing, to obtain the affection of one's carriers.[20]

While it is important not to overlook the fact that engineers in this period also expressed strong opinions about the inferiority of the white, unskilled workers employed in domestic construction work – 'the notoriously drunk navvy'[21] – there is no question that the perceived lack of technical ability of 'the native races' added to the drama of construction projects and by implication to the achievements of the explorer-engineer.

A connected and equally important element was that the writings of engineers *sensationalized* the surrounding environments in which the construction projects were set. Sensationalizing is a 'tag' introduced by Anne Crozier in a study of the writings of colonial doctors in Africa to denote that the characteristics of the tropical field were exaggerated with the purpose of creating an ideological effect.[22] Sensationalizing the African field by highlighting its abnormality, 'tropicality' and danger was a way for engineers to stress individual heroism and skill, and it enabled them to operate above usual social, moral and professional roles. The African field was one of wild animals, deadly diseases and hostile natives that only the *sui generic* explorer-engineer was fit to handle. The sensationalizing of the environment was another feature to set apart the explorer-engineers from earlier 'heroic engineers' whose achievements were set in a domestic context.

A number of large-scale engineering projects carried out in Africa in this period provided engineers with ample opportunities for presenting to the British public their talents as explorer-engineers. Illustrative in this respect was the Zambezi Bridge constructed across the Zambezi Gorge to the design of Douglas Fox & Partners with Charles Metcalfe as chief consulting engineer.[23] Construction of the bridge began in 1904 with the immediate purpose of establishing a connection to the zinc and lead deposits discovered in 1902 at Broken Hill in the northern part of Rhodesia.[24] No scaffolding was used during construction but instead a novel technique was employed in which steel wires were bolted into the banks in order to sustain the weight of two levers that were linked in mid-air. The bridge spanned the gorge in one single 500-foot arch and, with the waters of the Zambezi running more than 300 feet below, it was the tallest bridge ever erected.[25]

The project was followed closely in the British media, where the edifice was widely celebrated as a great triumph of British engineering, imperialism and ingenuity. Several factors made this engineering feat particularly capable of arousing public interest. Firstly, the location of the bridge by the Victoria Falls was a feature central to the public profile of the scheme; the waterfalls that Livingstone had 'discovered' and named after the British monarch in 1855 formed a spectacular and mythologized setting to the imperial engineering edifice and the connection with the eminent explorer could be taken as tangible evidence that engineers in their work were following his path and example. This played into a second issue, namely that the bridge was heralded as a link of the Cape–Cairo

Railway: the imperial vision of creating an all-British 'Iron Spine of Africa' from which the ribs of commerce and civilization would branch out, bringing light to the Dark Continent from one end to the other.[26] This railway vision was attributed to Cecil Rhodes but also to the engineers who kept the dream alive after the death of Rhodes in 1902. In particular, Charles Metcalfe was celebrated for selecting the site of the bridge next to the falls to honour a promise to his friend Rhodes, who allegedly had wanted the passengers to feel the spray of the falls as the trains steamed across the bridge on their way to Cairo.[27]

Thirdly, the design of the edifice was in a literal sense spectacular and the project became an icon in the flourishing visual culture of the British empire.[28] Natalia Starostina has argued that in this period photographic images of exceptional railway bridges in the colonial world were a seminal component in representations of the civilizing missions of the European empires and that such widely circulated photographic representations were instrumental in convincing large audiences in Europe about the superiority of Western technology and civilization over those of the colonial worlds.[29] It is difficult to measure the extent to which this was the case, but certainly the construction of the Zambezi Bridge was visually dramatic and offered explorer-engineers plenty of opportunities for displaying manliness and skill and for having them recorded for public consumption. The construction process, the bridge builders and the finished edifice were caught on camera and the photographs accompanied numerous articles on the project in technical as well as non-technical publications.[30]

An electric cableway in use at the construction site was a favourite motif. It became known in Britain as 'Blondil' after the French tightrope walker. It was used to carry materials and personnel from one side of the gorge to the other until the two cantilevers were connected. In his first autobiography published in 1904, Francis Fox included a photo and a letter from his son Charles Beresford Fox who at the time was resident engineer on the project site. In the letter the young explorer-engineer explained that he had carried the heavy cable for Blondil across the gorge using a temporary ropeway and described how he had experienced his first ride across with the Zambezi River far below:

> As they tied me into the 'Bosun's chair' I must admit to feeling a bit strange in relying absolutely on my own calculations for my safety. The chair is a piece of wood suspended by four ropes, with a canvas back and a sack and board as foot rest. Of course one is so tied in that were you to lose consciousness you could not fall out; this precaution, for some people, is advisable.[31]

By including in the autobiography the letter and the accompanying photograph of his son in the 'Bosun's chair' above the Zambezi, Francis Fox added yet another specimen to the many public descriptions of the works carried out by the falls. In addition to articles and interviews these descriptions included numerous postcards and souvenir lettercards featuring the construction site and Blondil. Such memorabilia were marketed from the very beginning and more were continuously added throughout the twentieth century.[32]

It was thus a well-known engineering feat that was inaugurated in due pomp and circumstance in Rhodesia in 1905. The opening ceremonies lasted three days and included celebratory speeches, formal dinners and sporting competitions. The inauguration also marked the culmination of the annual conference of the British Association for the Advancement of Science which was held in South Africa that year. The visit of the British Association was organized by the South African Association for the Advancement of Science, an organization that had been established only two years earlier in 1903 and modelled on the British association.[33] Metcalfe, who was a founding member of the South African Association and its president in 1904, had arranged that the inauguration of the bridge was included in the programme for the visit of the honourable members of the British parent association.[34] At the official opening ceremony, the inauguration speech was delivered on the bridge by the president of the British association, George Darwin, who quoted a forecast made in 1795 by his great-grandfather, Erasmus Darwin: 'Soon shall thy arm unconquered steam, afar / Urge the slow barge and draw the flying car'.[35] The sporting competitions included a day of racing, a day of athletic sports and a regatta on the Zambezi. The Rhodesian crews consisted of engineers and contractors' staff who rowed against teams that had travelled from the Cape Colony.[36]

The builders and engineers thus made the most of the potential that the bridge had for catching the attention of the public in Britain. During the time of construction and at the inauguration the edifice was used as tangible evidence of the solid and adventurous work carried out in Africa by British engineers. Indeed, already from its inception the bridge was so skilfully constructed as an icon of the achievements of British explorer-engineers in Africa that this view to some extent remains today.

Assistants, Chiefs and Consultants in the Public Eye

By the closing decades of the nineteenth century the explorer-engineer had become a cultural icon that embodied a recognizable set of characteristics and character traits that engineering in the colonial field was associated with and that engineers actively associated themselves with. Analysing this phenomenon further it is imperative to recognize that there was no uniform class of engineers and no single culture of British engineering to blend with the cultures of empire and exploration. Imperial engineers engaged with the public in a number of ways and for a number of reasons depending not least on their position within the profession. Beginning with a study of an assistant resident engineer who rose to public fame and from there moving to a chief engineer based in Africa before ending with a London-based consultant, this section analyses how the socio-professional hierarchy in the engineering profession shaped public activities and profiles among Britain's imperial engineers.[37] The final section of the chapter is devoted to exploring further the extensive public activities of the Great George Street Clique engineers who occupied the position at the very top of the hierarchy.

The Assistant

From 1898 to 1899 an Anglo-Irish Royal Engineer, John Henry Patterson, was employed as assistant engineer on the Uganda Railway. While working on the railway Patterson killed two lions and won fame as the slayer of the 'Man-Eaters of Tsavo'. The basic facts of the story of this explorer-engineer and his encounters with the lions are well known and can be recapitulated in brief; from March to December 1898 two male lions attacked and killed 27 Indian workers and an unknown number of Africans who were constructing a railway bridge by the Tsavo River in British East Africa. Following an intense and hazardous hunt the lions were eventually shot and killed by Patterson.[38]

Patterson had first published the story of the man-eaters in early 1899 in two short articles in the popular journal the *Field* and in 1907 he expanded his story of the lion hunt into an illustrated book titled *The Man-Eaters of Tsavo and Other East African Adventures*. The book is an instructive example of how railway construction was placed in the context of sensationalized Africa. The stories of game hunting and close encounters with lions, of 'superstitious Indian coolies' and of 'lazy Wakambas', run parallel with descriptions of plate-laying and bridge construction. The narrative is supported by photographs of railway work placed alongside trophy shoots of lions, rhinoceros and dancing 'natives'. According to the book's 1907 preface the enthusiastic big-game-hunter Theodore Roosevelt considered 'the incident of the Uganda man-eating lions ... the most remarkable account of which we have any record',[39] and one of its English reviewers claimed that 'in all books of adventure written since the days of Herodotus it would be hard to parallel such a story'.[40] Patterson's book was an immediate success and has remained so ever since, having gone through at least nine editions.

Contrary to what Patterson (and several writers after him) argued, the lions had no significant impact on the progress of the construction of the Uganda Railway (though it was a very serious matter for the workers that were attacked and sometimes killed). The works disturbed by the lions during two short intervals in 1898 were carried out to construct a permanent bridge at Tsavo in replacement of a temporary bridge that was already completed. The main workforce was occupied at the railhead which at the time was pushed forward from a point 40 miles further inland.[41] However, while the Tsavo lions had minimal impact on the ground in East Africa, they certainly conquered the public in Britain (and in North America) where the story gained almost instant fame. Over the following years Patterson and his lions were subjects of articles in youth magazines, travel literature and newspapers, and have since appeared in novels, tourist brochures and in three separate motion pictures including the 1996 Hollywood blockbuster *The Ghost and the Darkness*.

For Patterson the encounter with the lions in 1898 led to a remarkable change in his career trajectory. Very little is known of Patterson's life prior to his employment on the Uganda Railway, by the time of which he was in his early thirties. Purportedly the son of an Irish clergyman, Patterson had been trained as a military engineer in India where he also married a young teacher in 1895. It was from India he was transferred to East Africa.[42] The diary he kept while stationed in Africa reveals that the salary he made as assistant resident engineer on the railway enabled him to send home £20 a month to his wife and new-born child.[43] The fame he won as the slayer of the Tsavo lions, however, proved to be his entry ticket into the high-society whirl. Upon his return to Britain (after soldiering in the South African War), the military engineer had become known as one of great game hunters of his day and enjoyed the company of notables such as the actor Sir Henry Irving and the daughter of the explorer Sir Samuel Baker. A few years after his return Patterson published his full-length, sensationalized best-seller on his adventures in East Africa. For a Royal Engineer approaching middle age with a family to provide for, publishing this book was undoubtedly a much-welcomed financial opportunity. The year the book was marketed Patterson was appointed to the high position of chief game warden in British East Africa. He later returned to soldiering and became a leading and controversial figure in the Zionist Movement.[44]

The Chief

Associating their work with sensationalized African fields opened new opportunities for engineers in the public realms. However, not all engineers actively seized upon these opportunities. In this respect George Whitehouse, who as chief engineer to the Uganda Railway was Patterson's direct boss in East Africa, provides a telling contrast to his subordinate's public vigour. Whitehouse displayed very little interest in making use of the opportunities for public exposure that he undoubtedly had as chief engineer to the infamous Uganda Railway during the seven years it took to construct. He did not publish any books or articles on his East African experiences, he never lectured in the Society of Arts, the Royal Geographical Society or in the ICE in spite of the fact that he was a member of all three learned associations and went to their meetings when he was in London.[45] The few interviews he gave to agents from Reuter's were sobering in comparison with most contemporary statements about the Uganda Railway.[46] Whitehouse appears not to have had a taste for public exposure and as an employee in government service with an annual salary of £1,200 he may also have lacked sufficient incentive for seeking it.[47]

Others, however, were more than willing to do this for him. In 1903 the *Graphic* – at the time the most successful direct competitor to *Illustrated London News* – attributed the entire railway to the chief engineer when the popular journal carried a short biographical sketch of Whitehouse under the headline 'The Constructor of the Uganda Railway'.[48] The *Evening News* explained that Whitehouse had been chosen for the post as engineer-in-chief to the project largely on account of his wide experience in railway construction:

> He had already built lines in Peru, Mexico, India and Natal, but when he reached Mombassa in 1895 it looked to most people that he had undertaken a perfectly hopeless task. The difficulties which lay before him were stupendous, and it was only his indomitable energy and pluck that carried him through.

From there the article attributed the entire construction of the railway and a great deal more to Whitehouse:

> The first problem to solve was that of labour. The Swahili, or coast men, are too lazy to work, the Wakamba, further in land are agriculturists by deputy, all the hard work being done by the women, the Masai warriors would not dream of demeaning themselves by manual labour, and so forth, so Sir George brought his labour from India to the tune of 15,000 men, built corn mills to feed them, condensing plant to give them drink, and hospitals for the sick, and finally completed his task with wonderfully small loss of life considering the conditions. Sir George indeed appears to work on a Kitchener plan, and is undoubtedly an organiser of first rank.[49]

The strong personification of the Uganda Railway with its chief engineer was fed through many sources. A noteworthy incidence occurred when George Whitehouse's brother, Benjamin Whitehouse, lectured in the Society of Arts in January 1902.[50] Benjamin Whitehouse was commander of the Royal Navy and in 1898–1900 had been in charge of a survey of the British section of Lake Victoria.[51] In the Society of Arts he spoke under the headline 'To the Victoria Nyanza by the Uganda Railway' and credited not only the railway to his brother. He also emphasized George Whitehouse's role in the survey he himself had been in charge of by claiming that 'the British half of Victoria Nyanza with all the islands known to exist in it had been surveyed under the superintendence of the chief engineer'.[52] In the ensuing discussion to the lecture, Sir Guildford Molesworth – who had visited Whitehouse in East Africa in 1899 as inspector-in-chief to the Uganda Railway Committee – insisted that he thought that:

> Commander Whitehouse had been too humble in speaking about the wonderful way in which his brother had carried out the important work of the Uganda railway. The work of constructing the railway presented unique difficulties, which no other railway had to encounter. The difficulties were extraordinary. There was unknown

country, then the newness of the staff, the want of water, the absence of all means of
animal transport owing to the mortality caused by tsetse fly, the having to establish
everything, the policy of having to push forward the railway at any cost and at any
sacrifice, so as to get through somehow or other. Then there had been the difficulty of
lions, the difficulties of jiggers, which had often caused men to lose their toes ... The
way in which these enormous difficulties had been surmounted by Whitehouse and
his staff was deserving of the very greatest credit.[53]

Molesworth's sensationalized descriptions were reproduced in several of the next
morning's papers. In its coverage of Benjamin Whitehouse's lecture the *Morning
Post* mainly focused on the fact that the session had been chaired by Henry Morton Stanley on the very day of the explorer's 61st birthday.[54] *The Times*, however,
commented on the words of praise directed at the chief engineer of the Uganda
Railway and noted that it echoed the eulogy that the founder of the Royal Niger
Company, George Goldie Taubman, had written on George Whitehouse in
the paper a few weeks earlier.[55] Goldie, who had met Whitehouse in Uganda in
1901, had also attributed the completion of the railway to 'the energy, organisa-
tion genius, and fertile inventiveness' of its chief engineer.[56]

The different levels of public activity displayed by Whitehouse and Patterson
underline an important point. Sometimes it was the case that engineers actively
took on the role of the explorer-engineer while in other instances it was a set of
characteristics that was assigned to them by agents in their vicinity. Like other
groups, engineers did not fully control how they were presented and perceived
among the public. Tellingly, when the *Liverpool Daily Chronicle* in 1903 published one of the abstemious interviews Whitehouse gave to Reuter's it was
placed alongside a cartoon depicting a venomous snake confronting a sweating,
overweight passenger on the notorious Uganda Railway as the train made its way
into the heart of Africa.[57] Indeed, for Whitehouse, who unlike Patterson displayed little interest in a sensationalizing public, it was difficult to escape being
cast as a lion-fighting, railway-constructing civilizer of the Dark Continent.

The Consultant

The position of assistant engineer Patterson differed from that of chief engineer Whitehouse. However, the distance to the world inhabited by gentlemen
consulting engineers in Westminster was even greater. It bears repetition that
the imperial activities of consulting engineers were tied to government offices,
gentlemen's clubs, learned institutions and the conversations that took place in
these locations. Moving in the highest metropolitan circles, consulting engineers
engaged in public discourses on empire that added more layers to the ideas of
engineers and empire than those connected with the sensationalized fields of the
explorer-engineers' Africa. An illustrative case in point is Frederic Shelford, the
London-based consulting engineer to the railways in the West African Crown

Colonies. Shelford was very active in public spheres in Britain, where he in several respects combined the characteristics of the explorer-engineer with those of a metropolitan gentleman engineer.

Frederic Shelford's public appearances were closely related to his company's activities in West Africa. As described in Chapter 3, the railways constructed in this region by Shelford & Son for the Crown Agents were subject to extensive controversies and complaints, with contemporaries openly criticizing Shelford's projects for exceeding (very generous) timeframes, for being overpriced and for being engineered to an inferior quality. Those who believed that Shelford only kept getting assignments with the Crown Agents because he was married to the daughter of the permanent under-secretary for the colonies could, indeed, argue a very convincing case.[58]

In spite of the poor reputation of Shelford & Son – or rather, because of it – Frederic Shelford was eagerly seeking to establish himself as an authority on questions pertaining to engineering, imperialism and Africa. In the first decade of the twentieth century his name frequently appeared in the technical and non-technical writings on African railway development. Shelford's first public appearance in this regard occurred in July 1899 in the sensationalist weekly *Graphic*. The background to this article was that Shelford had led a survey expedition in unmapped territory between the coastal settlement of Sekondi and the Ashanti capital Kumasi with the aim of settling the trajectory for a future railway. The *Graphic* described the trials that the three white men of the expedition had surmounted and the accompanying illustrations were based on a number of sketches that Shelford had produced during the expedition and afterwards presented to the editors of the journal in London.[59] One sketch pictured Shelford and the white men swimming across a turbulent river, another showed native porters fleeing into the jungle, and a third sketch commemorated the arrival of the expedition at Kumasi. Such active use of visual representations was characteristic of Shelford's public appearances in the years that followed. He contributed more sketches and photographs to the *Graphic*, and in the numerous articles on the West African railways that were published in the technical journals it was common to see Shelford by a bridge, under a railway viaduct, outside station buildings, or floating in a ropeway above the West African rainforest wearing a gentleman's tie.[60]

Journalists were, however, more likely to find Shelford in London than in Africa. He was the first person interviewed in a series of articles that the journal *African Engineering* began in 1905 under the headline 'Talks with Engineers'. The series was – as the editor Stafford Ransome declared – 'devoted to engineers engaged in opening new regions [of Africa] for business'.[61] Shelford was interviewed in Westminster where the journalist had called upon the engineer at work in his office in Great George Street. In the interview Shelford explained that his interest in West Africa had first been aroused from reading *With Edged Tools* – Henry Seton Merriman's best-selling novel from 1894 set in West Africa – and he made clear how the railway lines that Shelford & Son constructed were rapidly opening the vast agricultural and mineral potential of the West African region for exploration.[62] The latter point usually featured prominently in Shelford's public writings and statements. When he was interviewed in his office by the *Financier* in 1901 'Mr Shelford was especially enthusiastic over the properties of *Ashanti Goldfields Corporation*, the resources of which he described as simply wonderful'.[63] He expressed similar enthusiasm with regard to other areas in West Africa in *The Times*, *Financial News*, *Mining Journal*, *African World* and numerous other newspapers and journals.[64] These public statements ranged from short extracts from interviews that Shelford gave to Reuter's to long reports from proceedings in learned societies and banquet dinners where Shelford addressed potential investors and fellow experts with African interests.[65]

In the latter case, the African Society constituted Shelford's most important institutional platform. Established in 1901 in honour of the memory and ideas of explorer and travel writer Mary Kingley, this society's objectives were 'the enhancement of knowledge pertaining to the law and customs' of West African societies.[66] Shelford was a founding member and in 1910 became its vice-president.[67] In Shelford's own account of the history of the African Society, he belonged to a group who 'wanted the Society to be Imperial' as opposed to other factions who 'wanted the Society to deal with Native affairs only, to be in fact, parochial and Pro-Native'.[68] On this ground Shelford protested against letting what he called 'arm-chair men of letters' (here he explicitly referred to E. D. Morel of the Congo Reform Movement) onto the council of the society.[69]

At the meetings of the society, Shelford addressed his audiences as an expert on colonial economics and politics. He was an outspoken champion of Chamberlain's ideas of a 'constructive imperialism', which stressed Britain's obligations 'to cultivate the undeveloped estates' of the empire for the prosperity of Britain and to the advantage of the inhabitants of the colonies.[70] In this view private trade and industry were the driving forces of economic development, but for these civilizing influences to flourish the establishment of infrastructural systems was needed and when necessary through public investments in harbours, telegraph lines and railway lines such as those designed by Shelford & Son.[71] The rostrum of the African Society was also used by Shelford to oppose vehemently what he referred to as the 'inaccurate articles' and 'perfectly absurd speeches' that criticized the West African railways constructed under his company's supervision.[72] Shelford claimed that it was a critique levelled by people who had little or no experience of the extreme conditions pertaining to engineering works in Africa. But, as he also insisted before a crowd of potential investors at the Liverpool Chamber of Commerce in 1900, he spoke as a man who had 'devoted both body and mind to the subject of West African railways for several years past' and he was therefore able to present the correct (and positive) assessment of the work done by British engineers in the region.[73]

Shelford not only claimed authority with regard to theoretical ideas of colonial development based on his West African experiences. He also successfully established himself as an expert on what he in 1907 labelled 'Pioneer Engineering' in a series of articles he wrote for *Engineer*.[74] The following year he expanded the articles into a book with the more broadly appealing title *Pioneering*.[75] The articles and the book described the equipment needed for engineering expeditions and survey parties, giving advice on a host of subjects ranging from the treatment of 'natives' to methods for running a traverse in dense tropical forests. These publications, Shelford explained, aimed not at 'the wealthy explorer who can spend what he likes' and who is looking for 'mere adventures' but at 'the pioneer who, in however humble a capacity, is doing good work, in introducing into undeveloped countries the advantages of civilisation and the benefits of twentieth-century talent and invention'.[76] Also in this case Shelford referred to his explorative endeavours and claimed that his 'hints to the traveller' were based on hard-won exploration experiences. He was a person who had 'swallowed a good deal of quinine', who knew from experience that 'Berkefeld's pump filters tended to clog up', and he noted that 'many a time has a soldier, or a civilian official examined my kit, and begged for an article or two from my equipment which has after many years' experience been brought to somewhat near perfection'.[77]

If long and frequent stays in West Africa were a prerequisite for knowing the needs of African colonies as well as those of 'the pioneer engineer' then Shelford does not qualify as an ideal expert. As far as can be established he only visited West Africa three times prior to 1910 and each time spent less than two months there. Like most consultants his real field of operation was the engineering offices in Westminster, the government offices in and around Whitehall, the lecture theatres of learned societies, the dinners of commercial bodies and the conversation rooms of gentlemen's clubs. One had in fact to be quite lucky to catch Shelford on a ropeway above the jungle in Sierra Leone. Yet, Shelford had the access, the need and the wish to make public use of his limited experience from the African field in pursuit of his agendas. In this regard he did well for himself.

In other respects and in the longer term Shelford, however, was much less successful. The frontier-going sons of the consultants in many instances filled the shoes of their illustrious fathers and went on to become leading figures in the engineering profession in their own right. Frederic Shelford was not among them. As described in Chapter 3, Shelford & Son was eventually ousted from the lucrative projects in West Africa. Moreover, unlike his father, William Shelford, the younger Shelford never was elected to positions of trust in the ICE

– the litmus test for professional status and social recognition within the civil engineering profession. With the right family background and a well-established imperial consultancy practice in Westminster, Frederic Shelford had the best starting point for penetrating and advancing in the ranks of Great George Street. He was also on his way to doing so. Born in 1871 he became a student member in 1891, was duly elected associate member in 1897, and transferred to the class of member in 1902.[78] In 1898 he organized a series of lectures on 'Railway Surveying in Tropical Forests' to which he also contributed a paper.[79] Yet, Shelford never made it into the ICE council where his father had been centrally positioned. Significantly, it was his business partner since 1907, Robert Elliott-Cooper, who was elected member of council in 1900 and who became ICE president in 1913.[80] Elliott-Cooper was also knighted for his services to the Crown, an honour also bestowed upon the two other engineers that Frederic Shelford had previously formed business partnerships with, his father Sir William and Sir Benjamin Baker. Even from a starting position in the inner track the way was not assured to the top of the profession in Westminster. It may well have been the case that Shelford's poor reputation and engineering skills prevented his advancement in the ranks of the profession. Certainly, it did nothing good to his standing in Great George Street when Shelford & Son's controversial dealings in West Africa in 1906 prompted a parliamentary survey on 'The Remuneration of the Consulting Engineers to Crown Colonies and Protectorates'. In connection with this survey a number of leading consultants, including major players such as Rendel & Robertson for the Uganda Railway and John Wolfe Barry & Partners for the Kowloon–Canton Railway in Hong Kong, were forced to inform the House of Commons on the exact amount of fees received for Crown Colonial projects during the previous decade.[81] After 1917 there is no sign that Shelford was active in the ICE and no obituary appeared in the institution's proceedings when he died in Johannesburg in 1943.[82] Tellingly, one has to turn to the journal of the African Society (which by then had become the Royal African Society) for an obituary notice explaining that Shelford in the 1920s had established himself as proprietor of Chisambo Tea Estate in Nyasaland, which became his home in Africa.[83]

Consulting Engineers as Public Intermediaries

Historians of technology have demonstrated that leading civil engineers of the early and mid-Victorian period were skilled at using the spectacles offered by large projects to promote their schemes to the wider public, as well as their personal characters and abilities.[84] Writing on the public strategies developed and employed by Marc and I. K. Brunel during the construction of the Thames Tunnel, Marsden and Smith sum up this point neatly by noting that 'engineers knew that their technological ventures were public – given meaning by publics, financed and supported by them, perhaps hindered by them and certainly conditioned by them'.[85] Public trust in the engineers was of seminal importance, in particular when vast financial layouts were involved in large infrastructural projects. This lesson was not lost to the engineers of later generations either, who, as in the case of the Zambezi Bridge, vested their projects with the imageries of the empire. Naturally, the imperial dimension was more palpable by the Victoria Falls in 1905 than it had been at the Thames in 1828 when the younger Brunel narrowly escaped premature death and won fame when water flooded the tunnel.[86] Yet, it is important to emphasize that during these decades the engineers whose projects and names circulated among the public were increasingly those who were involved in projects in the British empire. A particularly active group in this respect was the Westminster consulting engineers. Their role as intermediaries between Britain and projects in Africa and elsewhere was also a public affair and their names and status made them good candidates for conveying attention and credibility to engineering schemes. Moreover, their privileged position in London and their high-profile names made available to the consultants public channels that other engineers did not have – ranging from highbrow journals and technical publications to the halls and rostrums of learned societies and professional engineering associations.

In serving this intermediary public role, one important channel for consulting engineers was constituted by the widely distributed technical periodicals. As described earlier the market for engineering periodicals boomed in this period. The names of the Great George Street consultants appeared with striking frequency in the engineering journals. In the issue of *Engineering* of 31 May 1901 – this is the 'standard' issue examined in Chapter 1 – Douglas Fox and Charles Metcalfe were interviewed for a lengthy piece on 'The Design of Rolling Stock for Rhodesian Railways'. The illustrated article devoted particular attention to the spectacular Train de Luxe that Lancaster Railway Carriage and Wagons Company was building for the BSAC based on the design of Sir Douglas and Sir Charles.[87] The issue also featured an article on 'The Kumasi Railway', at that time under construction in the Gold Coast. The adjacent pages in maps, text and photographs made plain the progress of the works that were carried out by Shelford & Son.[88]

This instance reflected a larger trend; imperial projects and work were exceptionally well covered in the periodicals and consulting engineers were actively endorsing this. The extensive coverage of the irrigation and dam projects on the Nile in the 1890s and 1900s constitutes telling case in point; in *Engineering* and *Engineer* articles were equipped with photographs from the construction sites that, as was explained, 'were kindly put at our disposal by Sir Benjamin Baker, showing the progress of the great irrigation works which are in the course of construction on the Nile'.[89] Baker had taken the photographs of the works during an inspection tour from 26 January to 13 February 1900 and, referring to the content of the article, *Engineer* also explained that 'we owe also to the courtesy of Sir Benjamin Baker the extracts from reports of the engineers' resident on the works'.[90] In most cases the articles published in the technical journals on projects carried out overseas relied heavily on engineers' reports. Only on rare occasions were journalists and reporters sent out and remunerated by the producers of the journals in London. There was clearly a correlation of interests between on the one hand engineers, with a need and wish for publicity, and on the other hand producers of journals and magazines eager to cover spectacular British engineering feats in costly-to-reach corners of the world. Consulting engineers had assistants stationed at the works for longer periods of time and occasionally themselves travelled back and forth between the overseas construction sites and central London from where the journals were edited and published. This gave consultants a privileged position for getting their contribution to the successful execution of projects presented to a wider audience and by implication also their contribution to the advancement and consolidation of Britain's imperial presence and expansion in Africa and elsewhere.

The pages of the most prestigious highbrow journals were also open to top consulting engineers of the Great George Street Clique. In some cases the articles they wrote in such journals described ongoing projects but the columns were also used to launch new engineering schemes, in particular in cases where projects required risk capital to get off the ground or if they posed severe engineering challenges. Large-scale engineering works in the African continent in this period often fitted this description. In 1883, for example, in the highbrow *Nineteenth Century*, John Fowler and Benjamin Baker co-authored an article in which they laid before the readers a proposal to construct a freshwater canal from Alexandria via Cairo to the Red Sea. This would be a British-built and British-controlled alternative to the French-built Suez Canal that stood in need of deepening.[91] The timing of the publication of this article was by no means coincidental. The freshwater canal was one of the schemes Fowler and Baker had worked on while employed as general engineering advisers to Khedive Ismail and, as noted earlier, the position of the two consultants in the country was faltering when the public works department in Egypt was strengthening in the wake of the British occupation. Fowler and Baker clearly aimed to secure their position in the region by revitalizing the idea of a freshwater canal and to convince potential supporters and investors of the project's feasibility, economic potential and political advantages. 'In conclusion' the two engineers argued:

> The question of the construction of an alternative Suez Canal viá Alexandria and Cairo resolves itself into this – Is it, in a rainless country like Egypt, preferable to construct a sweet-water canal running along a ridge, or to widen a salt-water 'ditch' lying down in a hollow; and is it, as regards our own country, preferable to have an

alternative route for ships through Egypt, remote from the present one and under our own control, or to be dependent wholly upon M. de Lesseps and his successors.[92]

The freshwater canal, which according to the engineers' estimates would cost a staggering £10–12 million, never materialized. Yet, the article shows that Fowler and Baker thought it important to sway public opinion in favour of the engineering scheme and, while nothing came of this particular project, the article and the debate it prompted ensured that the names of Fowler and Baker circulated among the British public.

Fowler and Baker were alert to the geopolitical implications of the canal project and in arguing their case they referred to the economic as well as strategic gains the British could expect in return for the investment. In an 1889 article in the highbrow periodical *Fortnightly Review*, Charles Metcalfe, argued along similar lines but in a different context.[93] The article was part of the public campaign launched by the amalgamated interests behind the BSAC to win support for the idea of granting the company extensive mineral and land rights in the scrambled-for territories north of the South African colonies and states.[94] It was written by request of Cecil Rhodes who, according to Metcalfe, had said that he was 'looked on with some distrust at home' and therefore considered the engineer a better candidate for presenting the argument in Britain.[95] In the article Metcalfe argued that the best way for Britain to pursue her humanitarian, political and commercial objectives in the region would be for the government to grant a Royal Charter to a company as this would enable the British to protect 'the native against his hereditary enemy the Boer', to fend off German ambitions in the region and to secure a valuable piece of land suitable for agriculture and mining without straining the Exchequer. In bringing these results about, the engineer argued, railways were needed:

> A word as to the way in which the countries within the British sphere of influence in South Africa should be civilized and developed. The chief means plainly is the iron way: this is the great civilizer, the great developing force of the nineteenth century.[96]

The article was the first Metcalfe wrote for wider public audiences but over the following decades he was a conspicuous figure in the media in Britain. In his mother country he was known as 'Cecil Rhodes's railway man' and his name was above all associated with the Cape–Cairo Railway. Metcalfe first argued in favour of the transcontinental railway in the article in 1889 and thereafter vigorously promoted the idea in articles, pamphlets, books, speeches in London societies and in interviews he gave from the railway frontiers of southern Africa.[97]

In Britain, Metcalfe was a recognized expert on the development of colonial Africa, which for him essentially meant railway construction. In late 1915 he lectured on 'Railway Development of Africa, Present and Future' in the Royal Geographical Society and explained how the previously secluded continent had been 'made accessible through the efforts of explorers, missionaries and shipping companies, private companies and their engineers' and that Africa thanks to their joint efforts soon 'will lie open to all, a smiling, prosperous, and civilized continent'.[98] In the ensuing discussion his business partner in London, Francis Fox, alluded to what had become Metcalfe's hallmark:

> We engineers generally have to select the route of our railways by sending a theodolite party with levels over the country through which the line is to pass, but Sir Charles has a peculiar faculty about him and can do this by walking or riding over the district. He thinks nothing of starting a walk of 200–300 miles and he invariably proves to have chosen the right route.[99]

As an example of the intermediary functions of consulting engineers between imperial projects and public spheres in Britain the case of Metcalfe is worthy of note. He spent most of his career in South Africa but retained a durable profile and presence among the British public where he straddled the gap between an explorer-engineer educated through the trials of the field and a gentleman, aristocratic consulting engineer confidently discussing very rudimentary ideas of colonial development and grand visions of transcontinental railways. Metcalfe left Africa in 1916 for retirement at his country estate near Godalming southeast of London.[100]

For independently practising engineers involved in large-scale projects across the empire personal and public exposure was imperative and their businesses benefited from and depended on their status in the public as well-known and trustworthy agents. Moreover, their incentive for maintaining a strong presence in the media was further enhanced by the fact that they were not allowed to advertise their services through commercial channels. Historians have yet to explore systematically this aspect in the public life of British engineers but it shaped profoundly the public activities of the consultants and therefore warrants attention in this context. By banning independently practising engineers from advertising and touting their professional services, the young engineering profession had copied the example of the older professions of medicine and law. According to professional ideals, the advice and consultancy that engineers offered should be kept above the world of commercial advertising. This ban was furthermore in line with equally entrenched gentlemanly values and ideals stressing economic disinterestedness and independence – values that were relevant to the top layers of the profession aspiring to the status of gentlemen.[101]

Significantly, during this period this restriction was self-imposed by the engineering profession and there were no formal laws in Britain to prevent engineers from advertising consultancy services. In 1907 the London lawyer W. Valentine

Ball discussed the implications of this in a highly informed way. He pointed out that:

> Touting and advertising offend against no Act of Parliament; and it is in this respect that, for good or ill, there is a great difference between the brotherhood of engineers and other learned professions. The General Medical Council, for example, exercises certain disciplinary powers over registered medical practitioners. If a doctor is guilty of infamous conduct in a professional respect, he may be deprived of the right to practice.[102]

The same also held true of the professions of dentistry and of law and among professionals engineers were exceptional for having no formal law that imposed sanctions on practitioners who advertised their services. The ban was, as Ball pointed out, part of an unwritten code and professional etiquette that guided the profession. He quoted Westminster consulting engineer and member of council of the ICE, Alexander B. W. Kennedy, who in 1903 had argued before a body of students to imprint this code and ethos:

> There is a very strong temptation to a young man conscious not only of his own merits and abilities, but conscious also that he wants to get married and to make money, and that as yet he is known to but a few people, to tout round for business. That is a thing which must not be. There is, unfortunately no defined rule, as in the legal and medical profession, against it; but everybody who does so will be sorry afterwards.[103]

No Acts of Parliaments prevented engineers from advertising their professional services. There were, however, ways of sanctioning the thousands of engineers who wore the authoritative stamp as members of the ICE in case they violated the professional code and ethos on this point. Sanctions were distributed from the council room in Great George Street, the centre of the informal network of Westminster consulting engineers. The ICE council reacted promptly in cases · where members advertised consultancy services and were thus active in reinforcing the ban. In instances where the ethical code was broken on this point the response was to send a letter of reprimand to the perpetrator strongly urging him to put a stop to this.[104] The ultimate penalty that the council had at its disposal was the expulsion of members. However, the letter of reprimand from the ICE headquarters sufficed in most cases and at regular intervals the council recorded that a letter had been received from a member or a group of members who (in the standard phrase used in the council minutes) 'expressed regret for having advertised professional services in the press'.[105] As in most of its dealings, the council preferred to exert control by informal means when possible and only rarely did it explicitly need to enforce the code that banned against advertising.[106] The gain members could make from advertising consultancy services was clearly outweighed by the disadvantages of falling out of favour in Great George Street – an impression which is strengthened by the fact that it is extremely difficult to find in contemporary media any advertisements from engineers touting consultancy services.

The self-imposed ban on advertising influenced the public activities of the consulting engineers in several ways. By depriving the consultants of one important avenue for public exposure it added to the attraction of making appearances in articles and interviews. Doing so offered welcome opportunities that consulting engineers literally could not buy. Moreover, the ban clearly favoured established names with alternative access to public channels. This was above all the consulting engineers at the top of the profession. They could speak their cases in learned societies, address potential investors at banquets and dinners, present their schemes and express their opinions in highbrow journals and magazines. Indeed, it was the well-known consultants who least needed commercial advertising as their names appeared in interviews and articles, and in newspapers when knighthoods were announced.

Yet, the concern to enforce the ban had deeper roots. To lend credibility and trustworthiness to engineering schemes the public voices of the consultants needed to be perceived as retaining a high degree of impartiality and independence. For consulting engineers it was imperative to maintain a posture aloof from the commercial marketplace when they approached the public and they were therefore eager to fashion themselves and their functions as those of unbiased, independent professionals. Commercial advertising and touting were a threat to the credibility of this public posture. The code against advertising therefore needed to be reinforced if the consultants were to serve efficiently in a public context their role as intermediaries between Britain and engineering projects in the colonies. Moreover, the ability of consulting engineers to perform this function in public contexts depended not only on the status enjoyed by the individual engineer but also on how trustworthy the engineering profession generally was perceived among the British public. For consulting engineers protecting the 'public status' of the profession via the code and ethos enforced from the council room of the ICE was also a way of protecting personal interests. In this respect it was very easy for Great George Street consultants to see the general well-being of the engineering profession and the private interests of consulting engineers as two sides of the same coin.

Conclusions

Writing on the illustrious mid-Victorian engineers Marsden and Smith have pointed out that 'behind the myth of the heroic individual there thus lies a culture of reading and writing engineering practitioners, keenly responsive to audience, to texts and to the possibilities of self-advertisement'.[107] This was also the case with respect to the 'heroic engineers' associated with the empire during the late-Victorian and Edwardian period.

Britain's imperial engineers took on many roles as they engaged with publics in Britain; from the manly explorer-engineers roaming the sensationalized African continent to gentleman engineers in London discussing the future of the engineering profession in an imperializing world. As a result of these activities a range of images of the imperial engineer in Africa was produced and disseminated in Britain. Often a familiar and important contrast was drawn between on the one hand progressive, white imperial engineers and on the other a sensationalized, passive and dark African continent whose inhabitants were characterized as backward and incapable of appreciating let alone handling Western technologies. Yet the case of the engineers also goes beyond the construction of 'otherness' and hierarchies along racial lines. There was great diversity in the public activities and postures of engineers that owed much to their social classes and professional roles. As was highlighted in the comparative analysis of the public activities of Patterson, Whitehouse and Shelford, the position of engineers within the socio-professional engineering hierarchy was fundamental to the ways in which they pursued the opportunities in the public sphere offered by their imperial connections.

Among engineers in this period consulting engineers were the most vigorous and visible in public spheres in Britain. The leading consultants possessed the means and incentive for engaging with the public in pursuit of their agendas. They had access to media as well as institutions and were involved in projects that could arouse public enthusiasm, as in the case of the Zambezi Bridge. In the public profile of this engineering feat, self-representations of explorer-engineers heightened by the drama of construction against the sensationalized background of the Victoria Falls were brought together to produce an image of the British engineer as an imperial civilizer of Africa. Even the moderately observant onlooker could hardly avoid the conclusion that British engineers and in particular the towering figures of the profession were agents of empire.

While the consulting engineers had a range of different motives for engaging with the public the underlying dynamics usually involved an element of trust-building. Whether the aim of their public interventions was to defend ongoing projects against critique or to launch new schemes, the personal standing as well as the perceived trustworthiness of consulting engineers as a professional group were pivotal factors. How successful consulting engineers were depended to large degree on how they appeared and were perceived in wider public contexts. The level of activity that they displayed in the public sphere and their concern to protect the 'public status' of engineers provides strong evidence of just how important this was for their functions. Indeed, their role in preventing practitioners from advertising is testament to a general obsessiveness that they entertained with respect to protecting the status of consulting engineers as independent professionals. As Arapostathis has shown, consulting engineers were for this reason also eager to prevent practising consultants from taking out patents and instead worked to fashion themselves as unbiased, impartial experts at arm's length from the world of trade and industry.[108] The status as independent advisers was at the core of the social and professional identities of the Westminster consulting engineers; in line with the professional functions they served as well as with the gentlemanly aspirations of this consciously elite segment of the profession.

This analysis sheds light on what consequences the growing association with the empire had for how engineers were viewed in wider Victorian culture. It suggests that British engineers gained significant public respect through their work in the empire. Indeed, while it makes no sense to speak of engineers experiencing anything like the late-Victorian obsession with explorers and soldiers, there is little doubt that the 'heroic engineer' was given a new lease of life that this cultural persona would not have had without the element of drama added by narratives of exploration and imperial rivalry. The contexts of sensationalized environments and peoples clearly set the explorer-engineer apart from the fading Smilesian engineer whose achievements were set in the domestic scene. Furthermore, the image of the self-reliant and individualistic engineer may have been more convincing and easier to uphold in relation to projects in the colonial world at a time when engineering work in the British Isles increasingly came to be associated with specialized teamwork, bureaucratic organizations and employment in industry and laboratories. The idea of the manly, independent engineer trained through practice and the trials of life thus found a temporary refuge in the empire.

6 VANDALS AND CIVILIZERS IN ASWAN AND LONDON

The engineers are not to be trusted with it; they see only one side of the question. Many of them would calmly submerge all the temples in Egypt for the sake of a better irrigation scheme; and when we hear of all the advantages offered by the submergence of Philæ, we are inclined to doubt whether this is only the thin end of the wedge, and to say, Timeo Danaos et Dona ferentes.

H. H. Statham (1894)[1]

In the 1890s British engineers devised a project to construct a dam on the Nile at Aswan. The rationale of the project was to improve agricultural productivity in the country by increasing the amount of water available for irrigation in Middle and Lower Egypt. Damming the Nile at Aswan, however, had the drawback that it would cause the submergence of the island and temples of Philae adored by scholars, artists and tourists alike. The pending destruction of Philae sparked a controversy in Britain with London and Egypt-based British engineers placed in the centre of events. This chapter analyses this controversy in detail and demonstrates that the clash over dam and temple brought out and reinforced two conflicting understandings of what 'civilizing obligations' the British had acquired when Egypt was occupied in 1882; the obligation to modernize and the obligation to preserve.[2] It argues, moreover, that these contrasting views were rooted in ideas of contemporary Britain and of ancient Egypt and much less in understandings of contemporary Egypt and the perceived needs of the population in the region. Yet, the outcome of the controversy directly affected long-term developments in the Nile Valley.

The engineers involved in the controversy included the Royal Engineers of the British administration in Egypt as well as the Westminster consulting engineers Benjamin Baker and John Fowler. During the Philae controversy Baker was adviser and consulting engineer to the British administrators in Egypt while Fowler, in retirement from his active engineering career, served as president of the Egypt Exploration Fund, the largest British society for Egyptology and at the forefront of the opposition against the Aswan project. The analysis focuses in particular on the two leading Westminster engineers and demonstrates how the controversy revealed reservations about the Great George Street Clique and about engineers as 'civilizers' and trustworthy agents of the British empire. David Cannadine's point that 'on the boundaries of empire much was revealed about the social structure of Britain itself'[3] is true in this case and the controversy over dam and temple in the Nile Valley sheds light on the position of engineers in late-Victorian elite culture as well as on conflicting ideas among engineers concerning the role of their profession in an imperial world.

Modern Egypt, Ancient Egypt and the Engineers

For millennia the physical landscape, everyday life and political structures in the Nile Valley have been shaped by human efforts to utilize the waters of the river to suit their needs and ends. In the nineteenth century these endeavours were envisioned and at times also carried out on an unprecedented scale. Reflecting the changing geopolitical strengths and commitments in the region the Nile projects were organized by khedives in Cairo, the French and by the British in particular after the occupation in 1882.[4] At the time of the occupation the Egyptian economy was in a state of disarray to the extent that it threatened the returns on the heavy investments made during the reign of Ismail by foreign and especially British investors. Hence, the restoration of the Egyptian economy was a main objective during the first decades of the British occupation, a task that was left to the administration under the leadership of Lord Cromer.[5]

In a study of the Cromer administration, Peter Cain has demonstrated that the British administrators viewed the economic crisis in Egypt in the 1880s as originating from a deeper moral failure in the country. The harsh Gladstonian budgetary regime that the administrators introduced was meant to replace what in their eyes was a depraved 'oriental system' devoid of financial and moral constraint and moderation. Khedive Ismail, known among the British as 'the amazing spendthrift', may have been removed but he was seen as a product rather than a cause of a morally rotten system. The British administrators showed a contempt for the contemporary 'oriental mind' and saw themselves as the bulwark against what they considered an almost inherent decay that had brought Egypt to the brink of bankruptcy, and in the process impoverished the country's population. The Egyptians needed to be protected against themselves.[6]

The new administration embarked on a programme of modernization based on the introduction of new industrial techniques, ideologies of progress, institutions and technical experts.[7] The nexus of technologies, ideologies and personnel nurtured the idea that in Egypt transformation was particularly pressing and promising and that the British had the muscle to modernize the country. The British rulers considered an increase in the production of cash crops – in particular cotton – a vital element in bringing Egyptian finances to a sounder footing. In a practically rainless country this required more efficient utilization of the Nile waters. The Egyptian public works department was therefore strengthened and, unlike other departments of the British administration in Egypt, it was well funded from the beginning.[8] From 1882 and over the next decades this department was dominated by a group of British engineers specializing in irrigation and who, like Cromer, had been brought in from India. As was discussed in Chapter 3 these Anglo-Indian irrigation engineers constituted an influential and self-confident group that left a lasting stamp on Nile Valley.[9] In total, eight British engineers were transferred from India. The most prominent were Colin Scott-Moncrieff, who headed the public works department from 1883 until 1892 when he was replaced by another of the British engineers transferred from India, William Garstin.[10] The most controversial and colourful of the Anglo-Indian engineers was William Willcocks. The son of a British captain in the Indian army, Willcocks was educated at Roorkee Engineering College in India, and had worked in hydraulic engineering on the Ganges. When he arrived in Egypt in 1883 it was the first time he had set foot outside India. He was later knighted for his contribution to the design of the Aswan Dam.[11] Also important in the Philae controversy was Justin Ross, an Anglo-Indian colonel of the Royal Engineers who was in charge of restoring basin irrigation systems in Upper Egypt in the 1880s.[12]

Significantly, the moral dimension that in the eyes of the British administrators underpinned the British occupation of Egypt was also pronounced in relation to irrigation. In *Modern Egypt*, the apologia for his rule in Egypt that Cromer wrote the year after he had left office in 1907, he argued that improved irrigation was the most important of the many boons the British had bestowed upon the Egyptian population. According to Cromer the irrigation engineers had done more than providing water for the country:

> The British engineer, in fact, unconsciously accomplished a feat which, in the eyes of a politician, is perhaps even more remarkable than that of controlling the refractory waters of the Nile. He justified Western methods to Eastern minds. He inculcated, in a manner which arrested and captivated even the blurred intellect and wayward imagination of the poor ignorant Egyptian fellah, the lesson that the usurer and the retailer of adulterated drinks are not the sole product of European civilisation; and inasmuch as he achieved this object, he deserves the gratitude not only of all intelligent Asiatics, but also of all Europeans – of the rulers of Algiers and of Tunis as well as those of India.[13]

Cromer was sanguine about irrigation and its civilizing potential. Yet, in this respect he was no match for the engineers of the public works department. In 1902 William Willcocks in the Khedival Geographical Society in Cairo spoke on 'Egypt Fifty Years Hence' and rhetorically asked:

> What will the Nile Valley appear like to the traveller of fifty years hence? Green it
> will surely be; but it will no longer be a beacon pointing to the permanent prosperity
> which the irrigation systems of the ancient world could confer on a country. It will
> be a beacon showing what modern irrigation and modern science can do to develop
> agricultural wealth. The giant works in progress and in contemplation will put their
> impress on the country with no light hand.[14]

Willcocks's words were highfalutin but nevertheless reminiscent of a general outlook among the engineers in the public works department who adhered to a vision of a modernized, green Egypt engineered through improved utilization of the country's life source, the River Nile. On the front page for his book *Egyptian Irrigation* published in 1899 Willcocks summed up the point: 'Without irrigation there could be no Egyptian people, certainly no civilisation in Egypt'.[15] According to Cromer and the British engineers in the public works department, modern civilization and modern irrigation advanced hand in hand.

Egypt as a field for modern irrigation was not the only fixation the British had with the region in this period. Indeed, at the metropolitan end of the imperial axis modern irrigation was not the dominant Egyptian theme. It is well known that nineteenth-century British culture was marked by a zest for ancient Egypt; the land of the Old Testament, of temples and of harems. This deep and long-running theme in Western cultural history was spun to new heights as more travellers, new media, Thomas Cook's steamers and imperial expansion drew Egypt still closer to Victorian cultural life.[16] In art, literature, architecture and exhibitions Egypt of old featured prominently. More often than not the image of ancient Egypt was idealized and distorted, satisfying a European taste for the exotic. As has been argued by Edward Said and his followers, this 'orientalist discourse' also served to establish in the eyes of Europeans a contrast to the perceived degenerate state of present-day (nineteenth-century) 'orientals' as well as to a modern, rational, scientific and superior Europe.[17]

Victorian engineers shared the appetite for ancient Egypt. An early example is I. K. Brunel, who adopted an Egyptian-style design for the Clifton Suspension Bridge he engineered for the Avon Gorge in Bristol in the 1830s.[18] Later British engineers also published books on ancient Egyptian engineering and it was a favourite topic for speeches in professional and non-professional societies, where historical links were drawn between the engineers of ancient Egypt and those of the modern British empire.[19] Unsurprisingly, engineers who had developed links with Egypt during their professional careers displayed a particular interest in the Egyptian past. The fine art collection of Sir John Hawkshaw, for example, contained several paintings in the orientalist tradition including *Israel in Egypt*, Edward Poynters's masterpiece from 1867,[20] while Sir John Fowler, as will be discussed below, for over a decade served as president of the largest British society of Egyptology.

The keen interest British engineers took in ancient Egypt is, however, best captured by the case of the obelisk known as 'Cleopatra's Needle'. This 68-foot obelisk with an estimated weight of 224 tons had been quarried near Luxor during the eighteenth dynasty, 1500 BC, and in 1877–8 it was transported from its resting place since Roman times in Alexandria to London where it was erected on the Thames Embankment (where it stands to this day).[21] British engineers were heavily involved in this venture. John Fowler, who in 1877 was general engineering adviser to Khedive Ismail, had secured from the Khedive the permission needed before the obelisk could be removed from Egypt and transported to England and up the Thames.[22] The transport of the obelisk was partly financed by the engineer and contractor John Dixon who in the early 1870s had designed the Gizeh Bridge in Cairo. While based in Cairo Dixon developed an interest in the ancient history of Egypt and was involved in archaeological excavation works at the Pyramids.[23] Besides his role in financing the scheme Dixon also co-designed a custom-built metal vessel, dubbed *The Cleopatra*, which carried the obelisk to London. The other designer of this vessel was the young Benjamin Baker, whose name thus initially became known to the wider public. The design of *The Cleopatra* was innovative. The vessel measured 92 feet in length and contained seven watertight bulkheads through which the obelisk was placed in horizontal position during the passage to Britain. A small cabin for the crew was located on top. The obelisk was encapsulated in the cylindrical, container-like vessel on land and then rolled into the sea from the beach in Alexandria. Once seaborne, the mast and rudder were added and the vessel attached to a steam tug that towed it off for London.[24] In the lecture hall of the ICE Baker later explained before his colleagues that the transportation of Cleopatra's Needle was conducted much faster and cheaper than what French engineers had been capable of in the 1830s when they transported the 'Luxor Obelisk' to Paris and erected it at the Place de la Concorde.[25]

The central role of the engineers in securing the monument was widely acknowledged among the British public where the enterprise was closely followed in the media from the launching of the vessel in Alexandria accompanied by cheers of 'hurrah for old England'.[26] The extensively covered voyage of the obelisk turned out eventful as *The Cleopatra* was lost by the towing boat in a storm in the Bay of Biscay on 14 October. Six sailors died in a vain attempt to rescue it.[27] The vessel, however, stayed afloat and was eventually hauled to London by another tugboat. The obelisk was erected on the Thames Embankment on 12 September 1878 before large crowds and under comprehensive press coverage.[28]

The case of Cleopatra's Needle shows that engineers had the financial means, the technical know-how as well as the privileged position within the circles of power to enable them to devise the political and technical solutions needed before the obelisk could be erected in Westminster. A genuine fascination with ancient Egypt, personal ambition, a well-developed sense for publicity and a struggle for recognition within elite circles of British society drove the undertaking. The result was another material manifestation of the empire, a stone's throw from Great George Street, adding another dimension to the engineering heart of the British empire. A plaque was designed for the obelisk to honour those responsible for securing London's newest monument; five of the nine people mentioned were engineers, including Fowler and Baker. The plaque commemorated that the obelisk that had been quarried at the zenith of the Egyptian empire under Thothmes III and moved to Alexandria during the reign of Caesar had found its appropriate resting place in London in the forty-second year of Queen Victoria.[29] The symbiosis of modern engineering, ancient Egypt and British imperialism was unmistakeable – as it also was in the photos and sketches in engineering journals and popular illustrated magazines of the obelisk encapsulated in the steel vessel that Baker and Dixon had designed. This alliance of engineering, Egyptology and imperialism was one of the things that came under severe strain during the Philae controversy.

The Aswan Scheme and the Philae Controversy

In the first decade after the occupation of Egypt the Anglo-Indian engineers in the public works department were mainly engaged in organizing the restoration of existing irrigation systems that during the 1870s had fallen into a state of disrepair. In particular, resources were allocated for the rebuilding of the two barrages at the apex of the Nile Delta.[30] The main repair works on the barrages were successfully completed by 1890 and had a positive effect on cotton production in Lower Egypt. City bankers began to look upon the economic potential of Egypt in a more positive light as cotton exports from the country almost doubled from 1888 to 1891.[31]

Encouraged by the success in Lower Egypt the attention of the engineers and administrators turned to Middle and Upper Egypt as a field for further agricultural development. In 1890 the public works department instigated a survey of the Nile Valley with the aim of selecting a site for a new large-scale irrigation project. The rationale behind this new irrigation scheme was to turn the region into a two-crop cultivation area by storing the Nile water during the season when the river was high and then utilizing it in spring to grow an extra summer crop. Extensive surveys in the Nile Valley were carried out under the direction of William Willcocks and in 1893 the public works department issued a report dealing with alternative ways of bringing the irrigation scheme to reality.[32] The report concluded that from an engineering perspective the best solution to the agricultural and economic needs of Egypt was to create an artificial water reservoir in the Nile Valley by constructing a dam at the first cataract at Aswan situated 400 miles south of Cairo. Geological surveys revealed that at Aswan the rock bed consisted of hard syenite that would provide a solid, natural foundation for a dam. Furthermore, at this section the river was exceptionally wide, shallow and dotted with natural islands which meant that water could be directed between the islands thus making it possible to work in dry conditions for seven months annually. By constructing a dam, the tail end of the annual Nile flood could be impounded and then gradually released, thereby ensuring a high level of water during the summer season thus enabling the cultivation of an extra cash crop in the Nile Valley between Aswan and Cairo.[33]

The site at Aswan was superior from an engineering and economic perspective but it had one disadvantage; the reservoir behind the dam would flood the small island of Philae where the Temple of Isis stood alongside other Greco-Ptolemaic monuments. In their official report dealing with the question of water storage, the engineers of the public works department expressed hope that a different site for the dam could be found. If not, however, it was maintained that the material progress of Egypt necessitated that the irrigation project should go ahead. The report suggested that financing should be procured for moving the temples from Philae and re-erecting them on Biggeh Island where they would be safe from the rising waters once the reservoir began to fill.[34]

The scheme to dam the Nile at Aswan was further sanctioned by an international 'Technical Commission on Reservoirs' that assembled in Egypt to assess the conclusions reached by the public works department. The decision to place an irrigation scheme before an international commission was unusual. It was motivated by the technical challenges posed by the project, the huge financial outlay involved, the precarious political mandate that British rule in Egypt was based on, and by the pressure exerted by archaeological societies in Britain who, as will be discussed below, were protesting against the possible submersion of Philae.[35] The international technical commission consisted of Auguste Boulé, *inspecteur général* of Ecolé Nationale des Ponts et Chausées in Paris, Giacomo Torricelli, professor of irrigation and agriculture at Naples University, and Benjamin Baker, privately practising consulting engineer in Queen Anne's Gate, London. A leading authority in hydraulic engineering, Baker was the obvious choice for a British member of the international commission. He was, moreover, a renowned figure at the zenith of his career. Recently knighted for his co-design of the Forth Bridge, he was, by the mid-1890s, regarded as second to none among British civil engineers.

The three commission members met at Cairo in February 1894 and examined four potential reservoir sites with the task to make 'a selection from among the different projects, which have been submitted, for the information of the Egyptian Government'.[36] The commissioners, however, failed to reach unanimity in recommending a location for the dam. They were split over the question of Philae. As a consequence two separate reports were published, one by Boulé and one by Baker and Torricelli. In his report Boulé stated that he would not associate himself

> with any proposal for submerging or even modifying in any manner whatever, the ancient buildings which are to be found on the island of Philæ for they seem to me nearly equal in value to those of the Acropolis of Athens ... and if I acted thus I was certain to be held in contempt not only by my countrymen, but by the public opinion of the whole of Europe.[37]

Boulé, moreover, refused to accept that the question of Philae was beyond the expertise of the engineers of the commission, a point that the head of the public works department had insisted upon:

> I was, it is true, frequently told, (my colleagues appeared to admit it, and Mr. Garstin, seemed to say it in his report), that the questions concerning the destruction of temples were not within the competence of engineers. I absolutely refused to admit this (see the proceedings of the 24th of March), for its admission would have lowered the position of engineers to that of masons, carpenters, and mechanics.[38]

Torricelli and Baker in their report, however, supported the Aswan scheme, maintaining that placing a dam anywhere else in the Nile Valley other than Aswan was 'practically impossible' on account of the immense financial and engineering obstacles involved, and they therefore 'unhesitatingly recommended the adoption of that site'. The two engineers refrained from dealing with the Philae issue in their report as this was 'a question on which the Government has not asked the Commission for an opinion'.[39] Instead an appendix contained a proposal put forth by Baker alone in which he suggested raising the Temple of Isis above the future waterline along with the island of Philae 'without deranging a single stone'. This operation, Baker claimed, could be carried out well within a sum allocated for the preservation of Philae in the first report of the public works department.

The engineers in the public works department declared their agreement with the British and Italian commission members. In an official note to the reports of the commissioners, William Garstin as head of the department concluded that

> This narrows the question to very small limits and leaves the Egyptian Government face to face with the problem as to what is to be done with Philæ. If the dam be made at Assuan the temple must either be raised, removed or submerged.[40]

The Philae question thus seemed closed. The necessities of modern Egypt took priority at the expense of the ancient temples: 'the advantages to the country outweigh the sentiment' as Garstin phrased it in his note to Willcocks's first report on the Aswan scheme.[41] This was the view of the engineers of the public works department and of the majority of the technical commission. Importantly, it was also a view shared by Cromer, who in a confidential report to the Foreign Office in London stated his support for the site at Aswan:

> Attempt has been made to calculate the material advantages to be derived from the construction of a reservoir. Probably, in matter of this sort, it is difficult to make any accurate forecast, but it is certain that the advantages to be derived both by the population and the Government, are very great. Sincerely hesitant, as I am in any way to touch the temple of Philæ, I cannot bring myself to think that Her Majesty's Government would be justified in obliging the people of Egypt to forego these advantages in order to preserve the Temple intact. The archæological interests involved should, in my opinion, yield to the material wants of the inhabitants of the country.[42]

Among the British administrators in Egypt the dam thus took priority. However, when Cromer dispatched the report to the Foreign Office in London it was in response to the fact that in Britain the question of the inundation of Philae was not closed.

During the spring of 1894 an influential opposition was rallied in the imperial metropolis against the Aswan Dam on the grounds of the damage the project would inflict on the island and temples. The opposition against the dam gained strength from the fact that Philae was renowned in Britain. Indeed, since the early nineteenth century European painters, travellers, scholars, writers and tourists had flocked to Philae Island and produced and brought back a range of descriptions that praised Philae's artistic, historic and natural beauty. In combination these representations assured that Philae featured prominently in Victorian literary and visual culture. The novelist and travel writer Amelia Edwards, in her 1877 bestseller *A Thousand Miles up the Nile*, explained that the temple of Philae:

> has been so often painted, so often photographed, that every stone of it, and the platform on which it stands, and the tufted palms that cluster about it, have been since childhood as familiar to our mind's eye as the Sphinx or the Pyramids.[43]

Philae was familiar territory and its threatened inundation caused an outcry in Britain. The opposition against the dam was spearheaded by two organizations devoted to the Egyptian past. One was the Egypt Exploration Fund (EEF), founded in 1882 by Edwards in collaboration with British Museum official Reginald Stuart Poole.[44] The EEF was the leading association for Egyptology in Britain at the time and had organized a number of successful excavation expeditions in the Nile Delta and elsewhere in Egypt. These activities were funded by subscriptions and donations that the EEF obtained from members drawn mainly from the higher circles of British society. The founding and enduring success of the EEF is a strong testament to the zest for ancient Egypt that pervaded in particular the upper classes of late-Victorian society.[45] The second organization that headed the London opposition against the Nile dam was the Society for the Preservation of the Monuments of Ancient Egypt (SPMAE). This association was founded in 1888 by a group of leading British archaeologists and artists. Edward Poynter, watercolourist in the orientalist tradition and later director of the Royal Academy in London, was the driving force and honorary secretary (in the beginning together with the Egyptologist Henry Wallis).[46] The purpose of the SPMAE was not excavation but preservation – a task that the founders of the society considered an integral part of Britain's obligations in Egypt. The first statute of the SPMAE emphasized that Britain, by its occupation of Egypt, had incurred upon herself a responsibility for its ancient monuments:

> We [the British] have accepted the task of maintaining order in Egypt. We are bound also to safeguard those material evidences of Egypt's former greatness, both because the country is under our tutelage and because they are the common property of the cultivated in all lands ... The Pharaohs of the Ancient Empire recognised their obligation to take watchful heed of the fabric of the Temples. A venerable document relates that as far back as the 4th Dynasty a Prince Hortolef held the office of 'Inspector of Temples of Egypt'. The honour of maintaining and perpetuating the office has devolved upon us.[47]

The wheels of history had turned, so the argument ran, leaving it in the hands of the British to protect the temples. It was the fulfilment of this obligation that the SPMAE wanted to assist, thus defining the aim of the organization to be:

> The preservation of the splendid and interesting remains of the lands of the Pharaohs by aiding the Egyptian Government, as far as the means and influence of the society will admit, to maintain them and save them from further ruin, to protect them from the depredation of the Arabs, and from injury done by Tourists and others, by means of doors and enclosures, and by the means of appointments of responsible guardians and inspectors.[48]

The SPMAE considered it very urgent that the British should step in to safeguard the 'material evidences of Egypt's former greatness' in particular against the 'modern' Egyptians who were regarded as incapable of protecting let alone appreciating the temples. As the joint honorary secretary, Henry Wallis, explained:

> If there is one trait in the character of the European which is perfectly inexplicable and unintelligible to the Oriental, it is his taking trouble to visit and explore the monuments of antiquity. For the Egyptian, the Persian, the Turk, these records of the past are absolutely devoid of interest, saving as quarries.[49]

The leaders of the SPMAE in London shared the outlook of the British administrators in Cairo that the Egyptians needed to be protected from themselves. Like the EEF, the SPMAE also counted members of British high society among its supporters. In 1890 the Prince of Wales had consented to become its patron and Lord Wharncliffe its president.[50] One of the first initiatives of the SPMAE was to contact the foreign secretary, at the time Lord Salisbury, and the Foreign Office agreed to forward proposals from the SPMAE to the British administrators in Cairo.[51]

Prior to the outbreak of the Philae controversy the British administration in Egypt and the SPMAE had negotiated solutions in relation to two points in the agenda of the organization. These were firstly to find and finance a superintendent in Egypt to be placed in charge of preservation of the Egyptian monuments, and secondly to protect the temples at Karnak from infiltrating Nile water.[52] However, in the annual reports of the SPMAE Poynter complained that the British administration in Egypt was 'very slow to move' and should always be 'pressed upon' by the SPMAE to take any initiatives to protect ancient monuments.[53] Friendlier relations existed between the SPMAE and a few individual members of the British administration in Egypt. Sir Colin Scott-Moncrieff, who headed the public works department in Egypt in 1882–91, was a member of the executive committee of the Society.[54] Another supporter was the Anglo-Indian engineer Justin Ross, who in 1890 was placed in charge of the preservation works at the Karnak temples.[55] Despite these personal connections the SPMAE as well as the EEF were decidedly critical of the commitment of the British administration to the protection of the Egyptian monuments for which Britain, in the eyes of the Egyptologists and artists, was now responsible.

The Egyptologists only became further convinced of the British administrators' lack of commitment in honouring this obligation when rumours reached the Egyptological circles in London that engineers were surveying the Nile Valley for a suitable site for a large water reservoir in Upper Egypt and that Aswan was the preferred location. The Egyptologists were informed of this in 1891 when Ross wrote from Egypt to Edwards in the EEF to warn that the planned irrigation project might result in the drowning of Philae Island. He explained that the head of the public works department, Scott-Moncrieff, had been 'alarmingly weak on the subject' and had brushed aside Ross's objections by stating that 'a bath in the winter and summer up to the capital would do no harm' to the temples.[56] Edwards informed Poynter on the matter and urged the SPMAE to join the EEF in submitting a memorial to the Foreign Office and to the administration in Egypt protesting against the possible submersion of Philae.[57] Poynter responded by writing to the Foreign Office, who informed the SPMAE that the government in London had reached an agreement with the administrators in Egypt that the entire water storage question be placed before an international technical commission (what eventually became the commission consisting of Boulé, Torricelli and Baker) once the survey of the Nile had been completed. Besides committing themselves to an international commission, the Foreign Office stated that:

> Salisbury adds that although he cannot at the present stage of the proceedings express
> any definite opinion as to the particular project which is likely to be adopted he hopes
> and believes that measures will be found to improve the water storage without having
> recourse to the plans which involve the submersion of Philae.[58]

In the EEF Edwards expressed 'a certain measure of relief in what you [Poynter]
tell me about Salisbury's reply ... but that it may well be, that the evil is only temporarily postponed' and she therefore urged the SPMAE to raise a public storm
against the scheme:

> I am sure we must place no reliance upon the archaeological scruples of any engineer-
> ing commission. The sacred interest of the fellah and the still more sacred interest of
> the bond-holders will certainly turn the scale, unless public odium is made undeni-
> ably manifest.[59]

The SPMAE decided not to follow Edwards's suggestion as Poynter worried that
'an agitation might fall flat and could not be got up a second time'. He instead
arranged 'that a draft memorial against the proposed submersion of Philae be
prepared, to be circulated for signature to eminent persons in England and foreign countries' and to 'have appointed an agent in Egypt, offered a small salary
to keep us informed as to this and other matters'.[60]

The Egyptologists were thus well prepared when the threat to Philae became
more pressing in 1894. When Willcocks's report was made public, the SPMAE
was informed by the paid agent in Cairo that a dam at Aswan was the preferred
solution to the irrigation scheme.[61] Poynter set in motion the planned campaign
against the dam project. In *The Times* he announced a public meeting of the
SPMAE and explained that 'we intend at our general meeting to raise a discussion on the best means to be taken to oppose the contemplated destruction of
this world-renowned spot'. He explained that alternatives to the Aswan site were
available and insisted that the technical commission appointed to give its recommendations was unqualified to do this job in spite of the engineering merits of
its members:

> 'Rien n'est sacré pour le sapeur', say the French, nor, we may add, to the enterpris-
> ing engineer when he has a favourite scheme to recommend; and where, as in Mr
> Willcock's scheme, the construction of the 'biggest thing' in dams ever known is
> involved, it is not likely he will be induced to abandon the opportunity of so glorious
> an achievement without considerable pressure from the public opinion.[62]

The meeting held in the Society of Antiquaries on 23 February – three days
before the international technical commission had even assembled in Cairo –
was the first initiative in an exceptionally well-coordinated campaign against the
Aswan scheme orchestrated by the SPMAE and the EEF. In London, Poynter
was kept informed of the developments in Egypt by the paid agent in Cairo as

well as by Europeans in Egypt who were opposed to the project. Wilfred Blunt, author and outspoken opponent of the British occupation of Egypt, wrote to the SPMAE that it was insincere for the proponents of the project to claim that the dam would elevate the material conditions of the impoverished Egyptian *fellaheen*. This was not the case because the cultivators did not own the land that the dam would bring under perennial irrigation. Those who would profit were – besides British engineers and contractors – the large landowners and speculators who charged rent on the lands cultivated by the peasants.[63]

Blunt's information found its way to the final pages of a pamphlet published by the SPMAE as part of the campaign against the Aswan scheme.[64] This campaign, however, concentrated not on the consequences for the Egyptian population but on 'the act of vandalism' that the submergence of the temples would be. In its pamphlet the SPMAE expressed a measure of respect for the engineers who worked to ensure that Egypt 'shall enjoy throughout the whole year, the advantages it at present receives only during the comparatively short season of the Nile flood' but insisted that placing the dam at Aswan was irreconcilable with something of greater importance to humanity:

> The [Philae] Island and its surroundings in their natural features alone, form a scene of remarkable beauty, in no small degree enhanced by the noble and picturesque building with which the Island is crowned. Not only can nothing in Egypt be compared with it, but it may be doubted whether throughout the world a spot could be found where beauty, imparted by art as well as by nature, is so singularly combined with objects of deep historic and scientific interests ... To devastate the Island of Philae is nothing less than to rudely tear an important leaf out of the sadly mutilated volume upon which almost alone, scholars and students can rely in their laborious endeavours to trace the origin and early history of our modern civilisation.[65]

The campaign of the SPMAE climaxed in June when an official memorial protesting against the submergence of Philae was sent to the Foreign Office and also to the British administrators in Cairo. The memorial was signed by dozens of British scholars and gentlemen and arrived at the Foreign Office on the same day as similar protests issued from learned societies in Germany, France, Russia and Belgium.[66] Thus, the Egyptologists had a national and international network in place allowing the campaigners to add further pressure on the government in London at a time when the British occupation in Egypt was a controversial issue for international diplomacy. The leading members of the EEF also signed the memorial and added extra pressure on the Foreign Office and the British administrators in Cairo by dispatching directly to the foreign secretary an additional 'Memorial Against the Submersion of Philae', making the case again that safeguarding monuments overruled the obligation to irrigate the Egyptian dependency.[67]

Civilizers and Vandals in the Public Eye

The Philae controversy was debated with intensity in the British media and among the public. The predicament between dam and temple was simple but the controversy involved broader issues. It was disputed whether the development of modern agricultural methods and modern economy in Egypt should be the overriding British priority even at the expense of ancient monuments. Some debates regarded the new scheme to utilize the Nile as an ample demonstration of the determination, muscle and commitment of Britain to civilize and advance Egypt while others saw the apparent willingness to disregard unique temples for the sake of cotton cultivation as a token of how unfit Britain was to take on this responsibility. Indeed, the controversy brought out in the open conflicting ideas about what 'civilizing obligations' Britain had assumed by occupying Egypt.

The engineering press argued in favour of the dam. Throughout 1894 *Engineer* covered the Philae controversy extensively in a lengthy series of articles under the title 'The Utilisation of the Nile'. In addition to covering in detail the technical side of the Aswan scheme, *Engineer* complained about 'the storm in the teacup' raised by the Egyptologists on account of 'the possibility of swamping for a few months a year, the site of the famous Temples of Philæ' and insisted that

> the protection of an ancient temple, no matter how venerable, from its antiquity, or how likely to be instructive as a link between past and present, must not be allowed to stand in the way of an enterprise which will materially advance the prosperity of the population of Egypt ... On the scientific utilisation of the supply of the Nile depends the material prosperity and wealth of Egypt, and it is well known that to this end, ever since the British occupation of the country began, the labours of our administrators have tended. Now that a further evolution is on the eve of inauguration and an attempt is being made to solve one of Egypt's difficulties, exuberant fanaticism would shackle the hands of our engineers – such men as Moncrieff and Willcocks – and afford grounds for both home and continental pessimists to cavil at the advantages to the country of our protectorate.[68]

The competing weekly *Engineering* echoed the opinion of *Engineer* and speculated that

> Possibly, if the members of the learned societies who write to the Times on the subject, lived in Egypt, the material benefit that the reservoir would procure, to be reckoned in millions sterling per annum, would moderate their enthusiasm for relics of the past glory of the country.[69]

The engineering journals did not, however, represent the dominant point of view. During the first half of 1894 the public pendulum swung in favour of the temples. The sensationalist, illustrated weekly *Graphic* carried a stylized drawing of Philae Island and noted that many would 'consider the destruction of

Philæ a lasting blot on the British occupation of Egypt'.[70] *Illustrated London News* thought that the submergence of Philae was bound to renew 'the complaint against the utilitarian spirit of the age' and echoed a growing chorus when it insisted that 'we cannot wish these [temples] should be swept away'.[71] In the regular column on 'Fine Art Gossip', the *Athenaeum* joined the opposition and expressed hope that 'this country will not be made an object of scorn to the civilised world by perpetrating such an act of vandalism'.[72] The *Builder*, an influential illustrated weekly for 'the architect, engineer, archæologist, constructor, sanitary reformer and art lover', claimed in a front-page editorial that while Philae 'may appear to engineers to be a sentimental objection, trivial in comparison with the great national benefit, which would be conferred by the construction of the reservoir at this spot: it is, however, hardly to be dismissed'. The editorial blamed in particular the engineering profession, noting that 'the fact seems to be that the majority of engineers are absolutely unable to understand the interest which people of rather wider education and sympathies attach to monuments of architectural beauty and archæological value'.[73]

A number of individuals also entered the heated debate. One was Justin Ross, who had recently retired from his position as inspector general of Egyptian irrigation and as superintendent of ancient monuments. In the literary highbrow *Academy*, he called upon the archaeological world to oppose 'the act of vandalism' that the drowning of Philae would be and warned that:

> It is not enough to say that a committee of three engineers from England, France and Italy has been appointed to study the question: They were not sent in the interest of art, but to study the stability of the great dam. I do not wish for a moment to suggest that these three eminent hydraulic engineers are themselves vandals. Yet it is well known that engineers, when swayed by the interests of their calling, do not take into consideration the art side of the question; and it is not to them that we would naturally turn when we wish to preserve a world-famous monument, but to men of taste and archaeological knowledge.[74]

Upon his return from Egypt Benjamin Baker entered the public debate when he wrote an article in support of the Aswan project for the highbrow monthly *Nineteenth Century*. In his article Baker restated the point from the report of the technical commission that the only site 'practically possible' for constructing the dam was at Aswan and that it was an absolute necessity to go ahead with the project as 'the direct and indirect returns to Egypt must be enormous, and that the condition of the cultivators will be vastly improved'. Baker urged the readers 'to understand the Egyptian way of looking at the question': in the eyes of the Egyptians who, as Baker explained on their behalf, 'care not a piastre about the ruins' and 'who are constantly being reminded that England is in Egypt, not for her own benefit, but in the interest of Egypt herself', it was difficult to see why Britain would show a greater concern for the temples than for 'the vast material benefits which even the most ignorant of the fellaheen know must result from the establishment of the Nile reservoir'. Baker fiercely objected to any accusations of vandalism directed against the engineers who were left with the task of bringing about the future prosperity of Egypt and went on to explain how the same engineers without much difficulty could save the temples simply by raising them above the future water table:

> The well-drilled garrison at Assouan would be delighted to work the elevating screws with military precision, and no doubt can be entertained as to the success of the operation. When raised, the ruins surely must be of greater interest to the intellectual tourist than before. Half the wonder and admiration excited by the monumental works of ancient Egypt arises from the magnitude of the masses handled and transported by the old Egyptians rather than from artistic merit. It would be in accord, therefore, with the spirit of the surroundings if English engineers raised tens of thousands of tons where the Egyptians raised hundreds.[75]

Baker's claim about the historical link between ancient Egyptian engineers and modern English engineers was not well received. In the following issue of *Nineteenth Century* Frank Dillon, watercolourist and founding member of the SPMAE, attacked Baker and the engineering profession he represented:

> Sir Benjamin Baker takes exception to the term Vandalism in connection with the proposed destruction of Philæ. It must be admitted that the comparison is hard upon the vandals, who after all, were simply barbarians let loose upon the world in search of loot; while the modern engineers, with all the advantages of education and culture, seem to think that the world was created solely as a field for their enterprise and for opportunities of gain.

Dillon was in particular affronted by Baker's proposal to elevate the temples. He insisted that while 'art and engineering have not always been divorced' the suggestion to raise the temples showed that the engineers were 'absolutely without the *religio loci*, so important an element in the appreciation of architecture'.[76] The idea of Baker as a civilizer was severely questioned during the controversy. In the *Builder* the editor linked its rejection of the Aswan scheme directly to Baker's lack of aesthetic sensibilities:

> Sir Benjamin Baker, who represents this country on the Engineering Commission now engaged in reporting on the subject, is a great engineer, of whose fame and genius, as such we are all proud; but the fact that he has lent himself to the erection of the monstrosity in progress at Wembley Park [station] does not say very much for the acuteness of his Æsthetic sensibilities, and that is why we are inclined to support the view that the Philæ scheme should be taken out of the categories of alternatives. The engineers are not to be trusted with it; they see only one side of the question. Many of them would calmly submerge all the temples in Egypt for the sake of a better irrigation scheme; and when we hear of all the advantages offered by the submergence of

Philæ, we are inclined to doubt whether this is only the thin end of the wedge, and to say, Timeo Danaos et Dona ferentes.[77]

The relatively few voices that were raised in support of the Aswan project were no match for the well-coordinated chorus that in the public sphere as well as in diplomatic circles insisted that the flooding of Philae was irreconcilable with Britain's responsibilities in Egypt. The promoters of the Aswan project won the battle on the ground in Egypt but they were losing the equally important scuffle fought in London where the Foreign Office was under severe pressure from educated opinion in Britain as well as from abroad. Speaking in the Royal Institution in London after the controversy had been settled, the former head of the public works department in Egypt, Scott-Moncrieff, noted that: 'It was not only the English who were indignant. For once, I fear, since we occupied Egypt in 1882, was educated opinion in England and France at one. Both alike insisted that Philae should not be drowned.'[78]

With the Aswan scheme hanging in the balance the British administrators and engineers in Egypt began to work behind the scenes to find a compromise that would settle the controversy. In June, Garstin contacted Poynter in the SPMAE and reached a confidential agreement with the most outspoken critic of the Aswan project. On behalf of the British administration Garstin granted that £50,000 be spent by the Egyptian government on making a survey of the future reservoir area.[79] The survey of the Nile Valley was carried out the following year. It was organized by a committee consisting of members of the EEF and engineers of the public works department.[80] Another member on the survey committee was the editor of the journal *Nature*, Norman Lockyer, who at the height of the controversy predicted that one of the outcomes would be a detailed survey of a future reservoir area. This led him to speculate that 'dam or no dam we shall be infinitely better off from the scientific point of view than we are now or should have been for the next century, if the question of the dam had not been raised'.[81]

In addition to funding the archaeological survey the British administrators and engineers were also forced to draw up a modified version of the Aswan project. The alteration to the design consisted of the height of the dam being lowered by 8 meters which meant that the main temples on Philae Island would stay above the waterline of the future reservoir.[82] The decision to make this modification was voted at a cabinet meeting in Cairo in June in what William Willcocks later described as 'a great moment of weakness'.[83] From an irrigation point of view the reduction was indeed consequential; the reduced height of the dam meant that the reservoir would only contain one third of the amount of water originally planned. A note on the modified scheme that the public works department dispatched to the Foreign Office in London expressed regret that the modified project did not fulfil the needs of the entire country north of Aswan, but this way, perhaps, Egypt 'would reap the advantages to be derived from a storage reservoir without sacrificing the interest of science and archaeology'.[84]

The campaigners against the dam expressed greater satisfaction with the compromise. At the annual assembly of the EEF, the president of the society announced that:

> Communications had been made to the foreign office, which had been transmitted to Egypt, and further inquiry had shown that the material necessities of Egypt and the claims of archaeology were capable of reconciliation. The temples of Philae were now under the protection of the civilised world.[85]

The details for the modified scheme were drawn up in 1895 in London by Willcocks in collaboration with Benjamin Baker, who was appointed consulting engineer to the Aswan Dam.[86] The reduced dam was completed in 1902 and the reservoir began to fill; the water drowned most of Philae Island but the main temples stayed above the water table.[87]

Conclusions

Britain's imperial system was a staging ground for diverse, fluctuating and often contradictory ideologies, not least with respect to questions of tradition and modernity. The controversy over the temple and the dam allows us to see these ideological tensions more clearly. The Philae controversy revolved around two conflicting understandings of Britain's obligations in Egypt in the wake of the occupation. The defenders of the dam – while admitting that Britain had a measure of responsibility in terms of preserving ancient temples – emphasized the obligation to increase agricultural production in Egypt for the benefit of the Egyptian economy, creditors in Britain and the glory of the British empire. The antagonists of the Aswan scheme admitted that Britain had obligations to improve the material condition of Egypt but maintained that the preservation of the venerable monuments should take precedence over an irrigation project that aimed to enhance cotton cultivation. The basic question was whether the mission of the British was to protect or to irrigate, to preserve or to modernize. Both groups argued that the mantle of human civilization had fallen upon the British, but it was by no means undisputed what kind of responsibility and priorities this entailed.

While advocates on both sides of the controversy claimed to represent the standpoint of modern, civilized Britain, this did not prevent them from disagreeing profoundly. One point of convergence among the voices in the controversy was, however, that it was up to the British and not the Egyptians to decide whether dam or temple should take precedence. It was widely assumed that the Egyptians in their modern-day 'depraved state' were incapable of appreciating the country's past, that they were equally incompetent in analysing the requirements of the present situation, and therefore also incapable of making decisions for the future. Underpinning this controversy among the British was an assured feeling that they knew Egypt better than the Egyptians. This line of analysis also serves to explain why the Egyptians were 'at a distance' in the controversy. Not only were no Egyptian voices heard in the negotiations between the Foreign Office, the administrators in Cairo and the Egyptological societies; in the public confrontations during the controversy the consequences for the Egyptians were, at best, a secondary issue. Those who wanted to preserve Philae, at least, had realized that ancient Egypt, not the population of modern Egypt, was the more likely candidate to influence decision-making processes in London in this period.

Engineers were placed in the midst of the controversy. Most of the engineers whose voices have survived in the historical records argued in favour of the Aswan Dam. Yet, there was no singular engineers' point of view during the controversy. Auguste Boulé, the French engineer of the international technical commission, strongly opposed the dam on the grounds of Philae. So did Justin Ross, the Anglo-Indian irrigation engineer who informed the Egyptologists in London on the threat to Philae as it developed and who publicly opposed the project designed by his former colleagues in Egypt. While Ross and Boulé agreed that the drowning of the temples would be an act of vandalism they argued differently about what authority and responsibility engineers had in the matter. Ross maintained that it was up 'to men of taste and archaeological knowledge' to decide upon the question of the temples but Boulé argued the opposite case. He refused to accept that the Philae question was beyond the competence of engineers because in his view this 'would have lowered the position of engineers to that of masons, carpenters, and mechanics'.

The complex position of engineers during the Philae controversy is, however, best captured in the case of John Fowler, Benjamin Baker's patron and business partner for more than three decades. By the irony of fate John Fowler was president of the EEF in 1887–98 and thus also during the Philae controversy. Historians have hitherto not paid attention to this side of Fowler's life, with the exception of Buchanan who has noted briefly that Fowler in his retirement 'hobnobbed regularly with the aristocracy and even entertained royalty as well as distinguished political and artistic figures'.[88] Fowler, in fact, did more than hobnobbing. He joined the EEF in 1885 and was two years later asked to become its president, a chair he with some hesitation accepted and kept until his death in 1898.[89] He was not involved with the daily running of the EEF but he took an active part in discussions of the finances of the society, recommended new members for it, and chaired its annual general meetings.[90] After the death of

Fowler, E. Moude Thompson of the British Museum and centrally positioned in the EEF asserted that 'The choice of Sir John Fowler to guide the destinies of the rising society was a happy one', and that

> his success in piloting the fortunes of the Egypt Exploration Fund for eleven years, during which it rose to be one of the most prosperous and most energetic of the archaeological societies of Great Britain, was a source of satisfaction to others besides himself.[91]

The decision to appoint John Fowler president of the EEF is indicative of the very thin line between contemporary opinions of engineers as civilizer and engineers as vandals. In 1885 the EEF considered that the distinguished engineer had the personal and professional qualities required to fill the presidential chair. On his side Fowler through his professional and personal connections with Egypt had developed a long-standing interest in the country's past. He was a religious man and his fascination with ancient Egypt was partly motivated by a concern for biblical excavation work and archaeology carried out in the land of the Exodus.[92] He furthermore expressed respect and admiration for the engineering skills of the ancient Egyptians, as for example in a lecture in 1893 at the Merchant Venturers' School in Bristol where he rated the Egyptian engineers 'at least equal to the best work of the present day' in terms of 'workmanship, statuary, mechanical engineering, and architecture'.[93] In addition, Fowler's association with the Egyptologists was, in the words of his hagiographer, also motivated 'by his genial eagerness to enter into those pleasures of life which are approved by the most distinguished and most cultural section of English society'[94] – an eagerness Fowler shared with other Victorian gentleman engineers.

During the Philae controversy Fowler was a man of compromises. In March 1894 Justin Ross informed the secretary of the EEF that the president of the fund was a long-term associate of Baker, the British member of the international technical commission. Patterson wrote to Fowler and urged the president to 'use all your influence to prevent the submersion or removal of the temples of Philae'.[95] Fowler consented to this but chose only to advise the EEF on the engineering issues involved in the controversy. In one instance he explained to the secretary of the society why the technical solutions devised by British engineers were superior to those preferred by French engineers.[96] This was not the kind of support the EEF had sought but Fowler's loyalty clearly was with Baker, his former business partner of more than thirty years. When Baker's controversial suggestion to put the elevation screws to the temples was first made public the secretary of the EEF wrote to Fowler asking whether this was seriously considered as a solution and what the possible consequences would be: 'What will be done about Philae? If the island is raised will that not alter the orientation? Also can it be raised without causing the fall of the buildings upon it?'[97] Fowler replied that Baker had reassured him of the feasibility of raising the temples and that he therefore considered this a good way to solve the problem.[98] This opinion he repeated at a special meeting of the EEF on Philae where he almost echoed Baker and explained:

> the glorious ruins in the Nile Valley were records of the resources and power in constructive works, and at the present day the resources and powers of the Moderns had reached a development which made the design and construction of the vast barrage proposed a possible and safe work. His [Fowler's] idea was to use some of these resources to preserve Philae. There would he felt sure, be no difficulty whatever in raising every temple in the island to any required level without disturbing a single stone.[99]

In the speech Fowler had also emphasized the absolute necessity of preserving Philae against destruction, but his words were not well received by all members. The secretary of the EEF afterwards had to reassure a representative of the American members of the society 'that the president's speech does not express the policy of the fund, only of himself and you will kindly explain that if you are questioned about it. There are many members of committee who feel very differently about the Philae question.'[100]

The only public voice in favour of Baker's proposal to elevate the temples thus, ironically, came from the president of the EEF. However, while Fowler in public backed his business partner and friend, he nevertheless worked behind the scenes to find compromises to the dilemma. At the height of the controversy he wrote an official but personal letter to the foreign secretary urging that a settlement of the Philae problem should be found. Contrary to what Baker had claimed, Fowler insisted that it was possible to construct a dam further to the south that 'in the opinion of engineers does not present insuperable difficulties'.[101] As a dam further south would be of less risk to archaeology, Fowler urged the foreign secretary to reconsider the adoption of the Aswan site.

A different compromise settled the Philae controversy in 1894 but what makes Fowler's letter significant is that he was attempting not only to strike a balance between temple and dam but also between long-standing personal commitments in the EEF and towards Baker.

The case of Fowler in the EEF also underlines the fact that leading engineers eagerly and often with some success sought to obtain a position among the cultural elites of British society. Country estates, seats in learned societies and memberships of clubs such as the Athenaeum appealed to the gentlemanly aspirations of the consulting engineers in London. The ambition 'to enter into those pleasures of life which are approved by the most distinguished and most cultural section of English society' was also evident in the case of Cleopatra's Needle where engineers eager to leave a stamp in the heart of imperial London in this instance had their names put to the public and historical record alongside the names of rulers of Egypt, of scholars and of royalty.

However, the cases of Fowler in the EEF and of the transport of the obelisk appear more ambiguous in light of the storm over Philae. This controversy revealed that other elite groups entertained specific reservation about engineers as cultured, civilized agents – considering them more fit to arrange for the transport of monuments to the higher metropolitan circles than to speak in those circles. Indeed, during the Philae controversy arguments were raised about the lack of cultural sentiment and finesse among engineers and especially in the British engineering profession; the disregard for the temples displayed by towering engineers like Sir Benjamin Baker only exposed this 'cultural deficit' more clearly.

This perceived lack of culture in the engineering profession turned out to be a great hindrance for the promoters of the Aswan dam. The storm that blew in the face of them, and which ultimately forced them to modify their scheme drastically, developed the strength it did because there was a profound distrust in the ability of engineers to make the right judgements in this case; distrust in their ability to weigh the consideration for the temples against the consideration for the dam. 'The engineers are not to be trusted with it', as H. H. Statham put it, and the point struck a chord in Britain. This was perhaps the strongest argument that the defenders of Philae had at their disposal. Significantly, they did not criticize and attack the technical skills of the engineers or bring into question their ability to construct a stable, functioning and profitable dam that would solve the irrigation needs of Egypt. Instead, they argued that what could not be trusted were the engineers' aesthetic sensibilities, their sense for the importance of higher culture and their ability to exert self-control as their technical skills made possible ever more ambitious ventures. There was a great level of trust in the technical and financial abilities of the engineers, but in terms of 'cultural capital' the profession appeared much less affluent.[102] In the controversy over the temple and the dam this proved to be a decisive drawback for the engineers.

In the *Nineteenth Century* Baker defended his proposal to put the elevation screws to the temples by insisting that this action was 'in accordance with the action of Parliament in this country in relation to railways, the construction of which constantly and inevitably involves interference with cherished objects and the destruction of the picturesque'.[103] Such comparisons were likely to add fuel to the fire rather than calm the waters. In response to Baker's comparison between railway construction in Britain and the dam in the Nile Valley the classical scholar J. P. Mahaffy retorted in the same journal that this exactly cut to the root of the problem: engineers showed as little enlightened interest in history and culture in Egypt as they had done in Britain: 'If Londoners were willing to pay the price for an extra supply of water', Mahaffy argued in analogy to Philae, engineers would 'gladly dam up the Thames, so as to submerge all its valley as far up as Oxford, including Magdalen College which lies close to the river'.[104] In the controversy that unfolded over Philae loomed large a broader Arnoldian cultural critique of the 'utilitarian spirit of the age', which the rise of the engineers to a position of affluence and influence was easily seen as tangible evidence of. Indeed, the derogatory statements made at the height of the controversy about the lack of culture in the engineering profession were based to a large extent on assessments of the changes that British society had undergone and was undergoing. The outcry against the destruction of Philae was loud because there was a sense among other British elite groups that the engineers were now doing on the boundaries of the empire what they had already been doing in the British Isles.

CONCLUSION

This book has analysed imperial identities, networks and diasporas among British engineers with African connections. It has identified a number of functions served by the engineers that made them important agents in the British empire and it has demonstrated that imperial factors exerted profound influences on the British engineering profession in the period 1875–1914. In the course of the book the imperial dimension has been fleshed out in the reading cultures that developed around the engineering profession and notably in the professional journals that tied groups of engineers in Britain and in the colonies together on multiple levels. The study has, moreover, demonstrated that imperial influences were decisive factors in moulding locational patterns among British-based segments of an engineering profession whose leading members were congregated in Westminster, the heart of imperial London. It has also been analysed how the powerful metropolitan engineering networks – the social engines in a profession in which patronage structures and strong ties of obligation and trust remained crucial – were adapted gradually to a situation where assignments and revenues increasingly were generated from imperial and colonial projects. Imperial influences – metropolitan and colonial in origin – also shaped the ICE and the role that the most important of the professions' accredited institutions was envisioned to fulfil. The point has also been established that the profession, through activities of individual engineers in diverse public contexts in Britain, became more closely associated with late-Victorian cultures of empire and exploration and that the status of engineers in British elite culture was influenced significantly by the projects that engineers were involved with in the empire far away from Britain.

Cumulatively, this demonstrates that the empire struck back firmly in the profession and that imperialism exerted a profound influence among civil engineers during these decades. Equally important, however, the preceding chapters also show that the complex developments in the engineering profession in the period cannot be subsumed under a fussy concept and grand narrative of 'empire'. The imperial dimension was not singular and monolithic nor was it exclusive and sole; thus while imperial identities among engineers became more manifest in this period they constantly meshed with professional and social identities; similarly, the engineering journals helped to create and nurture a sense of inclusion in a range of different and often conflicting imperial 'projects'. And even in the ICE at the turn of the twentieth century the biennial conferences were organized to serve regional and national as well as imperial agendas. The book has unravelled and analysed the complexities of these issues and particularly the ways in which imperial impulses fed back into the profession. Indeed, while Angus Buchanan and others long ago noted the imperial dimension in the diaspora of British engineering, the feedback processes and the reciprocity of the imperial engineering connections have previously not been teased out adequately. In the case of engineers, the empire was not only something 'out there'. Therefore, rather than treating the imperial dimension in a separate compartment in the history of the British engineering profession, it is fruitful – in many cases necessary – to view Britain and the empire as a connected unity in which engineers were situated, constituting a seminal component in the geographical as well as the ideological context in which their activities took place.

A central concern throughout this book has therefore been to reshuffle spatial categories in order to analyse hitherto neglected or overlooked aspects of the activities of the engineers. In particular, reciprocal developments in Westminster and in various regions in Africa have been explored in a singular analytical framework emphasizing Britain and the empire as constituting interconnected zones. Within this framework the book has analysed the activities of British engineers based in Africa: the founding of advisory committees in southern Africa from the early 1890s, the work of the Anglo-Indian Royal Engineers engaged in barrage restoration and dam construction on the Nile, the agendas of the engineers and administrators in Nigeria who pushed forward at breakneck speed the first 'locally' built railway in West Africa in order to expel London-based consulting engineers from future railway projects in the colony.

The interconnected framework has also brought into focus the experiences of the 'go-betweeners', those who literally engineered for themselves 'imperial careers'. George Whitehouse and Charles O. Burge are illustrative examples of professionals who spent their working lives engaged for longer periods in engineering projects in various regions of the empire while retaining connections with Westminster. Another case in point is Stafford Ramsome, the engineer-turned-journalist who travelled widely across the empire, reporting back to Britain from shifting engineering frontiers, before he established the successful monthly *African Engineering* edited from offices in London.

Within the framework has also been set the analyses of the diverse activities and agendas of the imperial consulting engineers of the Great George Street Clique: of the streets they inhabited and the headquarters they constructed; of the networks their business were built from and the imperial connections they developed; of the institutions they dominated and the conferences they organized; and of the public profiles they entertained, the ancient relics they transported and the controversies they were involved in. In some of these cases the springs of developments mainly stemmed from engineers in colonial diasporas, in other cases from those based in Westminster, but what the cases have common is that British and 'colonial' developments were intertwined. They therefore must be placed within a singular analytical framework if sense is to be made of them.

The locational focal point has been London and specifically Westminster, which constituted an engineering centre from where enduring and consequential imperial connections were forged. The primacy of this location does not imply that all engineering connections went through London nor that the British imperial engineering world fits a simplistic 'spokes in the wheel model' in which all connections are radial, linking so-called 'peripheries' with the 'core'.[1] Also among engineers there were important cross-colonial linkages that did not bypass London and there were several other influential engineering centres in the British Isles outside the capital.[2] However, as the analysis has established, London was a crucial node through which engineers moved, contacts and contracts for projects were made and professional information and knowledge passed. This was a place where far-reaching connections were forged in particular for infrastructural projects that required backing and (occasionally financing) from the political system, which often was the situation with respect to the infrastructural projects in Africa that this study has been particularly concerned with. The primacy of Westminster as an imperial engineering centre was also recognized by the engineers who were drawn to and from Westminster in the course of their imperial careers as well as by the engineers who in this period clustered in the expensive district alongside the prestigious accredited institutions of the profession.

Those who benefited most from Westminster's unrivalled status as an imperial engineering hub were the consulting engineers of the Great George Street Clique. The imperial engineering connections were in most cases established by this elite segment of the profession. The consultants served a range of intermediary functions between on the one hand engineering projects in Africa and elsewhere in the empire and on the other hand groups and agents in Britain in the worlds of politics, finance, contracting and media. The imperial interests of the consulting engineers were widespread and their influence substantial. The Great George Street Clique is therefore a paradigmatic example of what John Darwin has referred to as the 'domestic' or 'second bridgehead' consisting of those 'enclaves of empire-minded or imperial-oriented interests in the metropole' and which:

Like the colonial bridgeheads ... was a composite of conflicting ideas, preoccupations and material concerns. Its members too had to find collaborators and struggle to exert their influence in an indifferent, occasionally hostile terrain. Sometimes they formed effective alliances with their overseas counterparts, sometimes not'.[3]

It was central to the professional functions of consulting engineers to make and propagate imperial connections both in Britain and in the colonies. Their strongest alliances and connections were in the metropole, as in the cases of Alexander Rendel with the Foreign Office, Douglas Fox & Partners (and Metcalfe) with the Exploring Company, Benjamin Baker with John Aird and Ernst Cassel and the Shelfords, the Coodes, the Hawkshaws and the Preeces (and a substantial number of other Westminster consulting engineers) with the office of the Crown Agents for the Colonies. Occasionally, they also established effective alliances overseas, as did Douglas Fox & Partners (and Metcalfe) with the South African mining magnates in the BSAC and Fowler and Baker with the khedival government and later with the Cromer administration.

For projects in Crown Colonies and protectorates metropolitan contacts were crucial to the consultants while in self-governing colonies and dominions strong connections at the colonial end of the imperial relationship were more decisive; a point that is evident, for example, in the case of Fowler's enduring and rewarding commitments in New South Wales that were established through his collaboration with John Whitton, his assistant (and brother-in-law) who migrated to New South Wales to take on the position as engineer-in-chief to the railways of that colony.[4] Yet, as was clear in the case of the ejection of Shelford & Son from the lucrative projects in West Africa, a failure to develop functional links and relations with British administrators and engineers based in Crown Colonies and protectorates could potentially undermine the consultant's otherwise secure metropolitan platforms.

The connections that the consulting engineers established and based their positions on were in most cases *systemic* in nature. In this context 'systemic' is a term introduced by Simon Potter to designate imperial connections that were excluding, monopolizing, stable and backed by institutions whose interests together 'dictate more formal, entrenched, and limited interconnection' and to differentiate these from other kinds of imperial connections that were open, flexible, dynamic, competing and subject to negotiation.[5] While welcoming a 'networked conceptualisation' of the British empire, Potter rightly emphasizes the necessity of not losing sight of institutionalized inequalities of power and he encourages us to pay particular attention to periods where imperial connections did become more systemic in nature. A tendency towards systematization of the imperial connections was notable in the case of consulting engineers. Their tightly knit networks were based on 'strong ties' and patronage structures (often underpinned by kinship), and the consultants preferred to operate and recruit through these networks to maximize trust as well as professional and social skills. The imperial connections they forged with other groups were kept within their networks which to some extent were monopolistic and clearly worked to the detriment of the opportunities of outsiders. Occasionally, this issue was commented upon by contemporaries and it was certainly also felt by engineers such as George Whitehouse when upon his return to Westminster after the completion of the Uganda Railway he was informed by Alexander Rendel that he would never get another job. The systemic nature of the consulting engineers' connections was further backed by their privileged access to institutions in Westminster: government offices, London clubs and above all the ICE, which was a powerful institutional platform for the informal network of consultants. It was from the council room of the ICE that the consultants enforced the ban on advertising that privileged well-established practitioners and businesses and it also gave them a position to settle the conditional terms on which connections with ICE members in colonial diasporas were established.

The systemic influence of the consulting engineers worked to reinforce London as an imperial engineering hub and to link the engineers – and the infra-structural projects they organized – more firmly with the established financial and political structures and institutions of power in London. This was a con-sequential development. As we have seen, the editor of *Engineering Magazine*, Charles B. Going, argued that in a world of global industrial imperialism tradi-tional leaders – statesmen and generals – would 'become servants and ministers of the engineer'.[6] In Going's vision the work of the engineer in an imperial world was a subversive force, changing the face of the earth and the power structures *within* industrial societies in the process. In the view from Going's editor's chair in 1899 the next century belonged to the engineer and the modernizing forces he represented. Yet, segments of the engineering profession – more influential than editors such as Charles Going – were more preoccupied with finding a niche in established metropolitan elite society than with subverting it.

The issue of what social and cultural position late-Victorian engineers held among British elites has been a central concern in this study. Buchanan has discussed this theme at some length and asserts that between 1750 and 1850 engineers 'won for themselves most of the trappings of social esteem, and from 1850 to 1914 were able to enjoy the full recognition of this status'.[7] Buchanan does not make clear social distinctions within the engineering profession and therefore leaves the impression that the whole profession had entered elite cir-cles. However, with regards to a small segment of the profession the assertion holds true. From the second half of the nineteenth century, leading engineers to an increasing extent wore the marks of elite society; fashionable clubs, learned associations and, occasionally, seats at country estates were open to the most prosperous individuals in the profession. They maintained an income and enjoyed a social and professional status that they used not only to establish a position as 'gentlemen engineers' but also to mould and reinterpret the category of 'gentlemanliness' to encompass their own activities.[8]

This book has demonstrated that the empire played a crucial role in this process. Imperial engineering was an effective social lever for British engineering elites during this period. For successful members of the profession the empire provided a path to affluence and wealth as well as to high positions in accredited institutions. It gave opportunities for engineers to obtain the honours and knighthoods which the execution of infrastructural projects in the empire was particularly likely to release, and provided, moreover, a platform for gaining professional recognition among peers as well as exposure in the wider public realm. This imperial lever, moreover, worked most efficiently in and around Westminster where the consulting engineers developed their connections with political and financial groups and institutions with imperial interests. This network of Westminster engineers nurtured and eagerly displayed commitments to the British empire as, for example, in the notable imperial identity and character of the ICE that the consulting engineers were instrumental in fostering during this period. To this group being 'the engineers of the empire' was a defining characteristic; a pivotal element in their professional and social identity and one of the ways in which this generation of leaders in British civil engineering differed from their early and mid-Victorian predecessors.

Thus the empire was clearly important to this group, but more must be said about what kind of imperial 'project' and idea these elite engineers adhered to. Thompson has suggested a useful three-way distinction and argues that among British elites there were at least three competing conceptions of the empire: an 'empire of privilege' (espoused by the aristocracy and landed gentry), an 'empire of profit' (espoused by entrepreneurs) and an 'empire of merit' (espoused by the professional middle classes). Thompson sensibly argues that engineers belong to the latter category.[9] Indeed, it was in many respects an 'empire of merit' that the consulting engineers adhered to. To them the empire meant a place where individual engineers could excel and gain professional recognition for ambitious engineering feats set in challenging physical environments not found in the British Isles. They, moreover, regarded the empire as the proper training ground where the next generation of engineers – and in particular their own next of kin – could learn the trade, earn merits and build the character traits needed for the successful continuation of engineering dynasties. And they saw the empire as a structure that provided a geographical, political and ideological setting in which the British engineering profession would continue to contribute to the grandeur and power of Britain and thereby allow for the continuation of the remarkable rise in the power of the engineering profession that was a manifest feature of the nineteenth century.

Consulting engineers thus championed an imperial 'project' that matched the ideals of the rising professional classes. However, they also forged lasting connections with and were influenced by groups that were in the business of promoting 'an empire of profit' such as mining companies, contractors and financiers. And while consulting engineers preferred not to speak about their road to affluence many of them became exceedingly wealthy from their imperial connections. Such connections were bound to create tensions with professional ideals that emphasized the status of consulting engineers as independent professionals. This was, indeed, a tension that pervaded the position of the consulting engineers of the empire who struggled to maintain credibility as impartial, gentlemen professionals. Moreover, as consulting engineers became part of the political and cultural establishment they also began to rub shoulders with the traditional elites that embraced an 'empire of privilege' or in Cannadine's terms an 'empire of tradition'.[10] As became evident during the Philae controversy this caused friction and brought out reservations about the status of engineers as civilized agents of empire. In this case other elite groups were concerned that engineers did not have the culture and formation required of a ruling elite. Having an eye only for professional potency and pecuniary interests, engineers were not seen as fit to be steering the course of the empire. Indeed (to argue along with Cannadine), the traditional elites who were inclined to see the empire as an extension of a hierarchical, traditional British society were certainly not willing to view it as a playing field for the execution of grand, modernizing engineering schemes.

The case of Philae shows that such groups could muster a powerful opposition against the engineers and ultimately force them and their associates to retreat from a prestigious project that otherwise enjoyed political backing. However, such protests against the imperial influence and connections of engineers must be placed in a proper context. The Philae controversy was not testament to a more general cultural outlook in Britain of industrial decline which – as has been argued – caused a loss of confidence among late-Victorian engineers.[11] Lack of confidence was not notable among the Great George Street Clique. On the contrary it was not least because of their continuous rise in power and influence and, perhaps, their lack of humility that these engineers got involved in controversies and disputes with other late-Victorian elites such as scientists, architects, reformers of technical education and – as in the case of Philae – scholars and literati. Indeed, there was no loss of self-confidence among the leading engineers of the late-Victorian and Edwardian generation who had their peers commemorated in Westminster Abbey where a window to honour Benjamin Baker was added in the north aisle in 1909. As Christine Macleod rightly asserts:

> The engineering profession was on the crest of the wave. What reason could there be for embarrassment or humility ... From Kelvin's transatlantic telegraph cable to Baker's Forth rail bridge and Aswan dam, it is hard to imagine that any Edwardian engineer could doubt that these were still 'glorious times'.[12]

The basis for this optimism stemmed not least from an assured feeling that the future of the British engineering profession was imperial. For the consulting engineers who planned, oversaw the construction and in 1913 inaugurated the new imperially ornamented ICE headquarters in Great George Street to house the growing membership and future generations of Britain's metropolitan and colonial engineers, the foreseeable future of the profession looked bright as well as imperial.

NOTES

The following abbreviations are used in the notes:

Advisory Committee for South Africa	Archive of the Institution of Civil Engineers, Advisory Committee for South Africa 1904–50, no. 189/107.
CMB	Archive of the Institution of Civil Engineers, Council Minute Books.
EES	Egypt Exploration Society Archive; material is cited EES followed by box number, envelope number (designated by capital letter) and document number.
PICE	*Proceedings of the Institution of Civil Engineers.*
'Reports of the Technical Commission'	Technical Commission on Reservoirs, 'Reports of the Technical Commission on Reservoirs with a note by W. E. Garstin', Archive of the Institution of Civil Engineers, no. 627.8(62) B.
Transvaal Correspondence	Archive of the Institution of Civil Engineers, Advisory Committee: Transvaal Correspondence 1893–7, no. 189, box 20.

Introduction

1. H. L. Wesselink, *Divide and Rule: The Partition of Africa 1880–1914* (London: Praeger, 1996); J. E. Flint, 'Britain and The Scramble for Africa', in R. Winks (ed.), *The Oxford History of the British Empire, Vol. V: Historiography* (Oxford: Oxford University Press, 1999), pp. 450–63.
2. T. J. Misa, *Leonardo to the Internet: Technology and Culture from the Renaissance to the Present* (Baltimore, MD: Johns Hopkins University Press, 2004), pp. 97–127.
3. R. E. Robinson, 'Introduction: Railway Imperialism', in C. Davis and K. Wilburn, with R. E. Robinson (eds), *Railway Imperialism* (New York: Greenwood Press, 1991), pp. 1–7, on p. 4.
4. R. Kubicek, 'British Expansion, Empire and Technological Change', in A. Porter, *The Oxford History of the British Empire, Vol. III: The Nineteenth Century* (Oxford: Oxford University Press, 1999), pp. 258–77, on p. 263.
5. R. A. Buchanan, 'The Diaspora of British Engineering', *Technology and Culture*, 27:3 (1986), pp. 501–24, on p. 503. See also R. A. Buchanan, *The Engineers: A History of the Engineering Profession in Britain 1750–1914* (London: Jessica Kingsley Publishers, 1989), pp. 146–61.

6.   G. Paish, 'Great Britain's Capital Investment in Other Lands', *Journal of the Royal Statistical Society*, 72:3 (1909), pp. 465–95. The total combined length of African railways was estimated to be 18,000 miles.

7.   The classic text for the perspective of 'benevolent diffusion' is G. Basalla, 'The Spread of Western Science', *Science*, 156:3775 (1967), pp. 611–22, which by implication also applies to technology. Much of the writing produced in the colonial period was also written in this assured vein. The classic text for 'oppressive diffusion' is D. R. Headrick, *The Tools of Empire: Technology and European Imperialism in the Nineteenth Century* (New York: Oxford University Press, 1981).

8.   For an excellent historiographical overview of this literature and the field, see D. Arnold, 'Europe, Technology and Colonialism in the 20th Century', *History and Technology*, 21:1 (2005), pp. 85–106. For a critique of diffusionist explanations and a persuasive study that demonstrates how technical knowledge and practices circulated in more complex ways than in transfers from the 'core' in Europe to the 'colonial peripheries', see D. Arnold, *Science, Technology and Medicine in Colonial India* (Cambridge: Cambridge University Press, 2000), pp. 92–128. See also W. K. Storey, *Guns, Race, and Power in Colonial South Africa* (Cambridge: Cambridge University Press 2008); D. Edgerton, *The Shock of the Old: Technology and Global History since 1900* (New York: Oxford University Press, 2006).

9.   A. Thompson, *Imperial Britain: The Empire in British Politics c. 1880–1932* (London: Longman, 2000); J. M. Mackenzie (ed.), *Imperialism and Popular Culture* (Manchester: Manchester University Press, 1986); A. Burton, *Burdens of History: British Feminists, Indian Women, and Imperial Culture, 1865–1915* (Chapel Hill, NC: University of North Carolina Press, 1994); R. Drayton, *Nature Government: Kew Gardens and the Improvement of the World* (London: Yale University Press, 2000). For an overview and critical discussion, see A. Thompson, *The Empire Strikes Back? The Impact of Imperialism on Britain from the mid-Nineteenth Century* (Harlow: Pearson Education, 2007).

10. B. Porter, *The Absent-Minded Imperialists: Empire, Society and Culture in Britain* (Oxford: Oxford University Press, 2004); and for a rejoinder in the debate, see J. M. Mackenzie, 'Comfort and Conviction: A Response to Bernard Porter', *Journal of Imperial and Commonwealth History*, 36:4 (2008), pp. 659–68.

11. Thompson, *The Empire Strikes Back?*, p. 5. This point is also argued persuasively in S. J. Potter, 'Empire, Cultures and Identities in Nineteenth- and Twentieth-Century Britain', *History Compass*, 5:1 (2007), pp. 51–71.

12. For Glasgow as an imperial city, see J. M. Mackenzie, '"The Second City of the Empire": Glasgow – Imperial Municipality', in F. Driver and D. Gilbert (eds), *Imperial Cities: Landscape, Display and Identities* (Manchester: Manchester University Press, 1999), pp. 215–37. For a study of imperial connections between Glasgow and Bombay which also involved engineers, see S. Hazareesingh, 'Interconnected Synchronicities: The Production of Bombay and Glasgow as Modern Global Ports c.1850–1880', *Journal of Global History*, 4:1 (2009), pp. 7–31. For a study of a Glasgow network that expanded its operations to Africa during this period, see J. Forbes-Monroe, *Maritime Enterprise and Empire: Sir William Mackinnon and His Business Network, 1823–1893* (London: Boydell Press, 2003).

13. Important studies in this historiographical reorientation include A. Lester, *Imperial Networks: Creating Identities in Nineteenth-Century South Africa and Britain* (London: Routledge, 2001); S. Dubow, *A Commonwealth of Knowledge: Science, Sensibility, and White South Africa 1820–2000* (Oxford: Oxford University Press, 2006); Z. Laidlaw, *Colonial Connections, 1815–45: Patronage, the Information Revolution and Colonial Government* (Manchester: Manchester University Press, 2005); T. Ballantyne, *Orientalism and Race: Aryanism in the British Empire* (Hampshire: Palgrave Macmillan, 2001).

For a good survey of the literature on networks and the British empire, see G. B. Magee and A. Thompson, *Empire and Globalisation: Networks of People, Goods and Capital in the British World, c.1850–1914* (Cambridge: Cambridge University Press, 2010), pp. 45–63. For an insightful overview that stresses the potential but also the limits of viewing the British empire in terms of network structures, see S. J. Potter, 'Webs, Networks, and Systems: Globalisation and the Mass Media in the Nineteenth- and Twentieth-Century British Empire', *Journal of British Studies*, 46:3 (2007), pp. 621–46.

14. The best synthesis is J. Darwin, *The Empire Project: The Rise and Fall of the British World-System, 1830–1970* (Cambridge: Cambridge University Press, 2009).

15. J. Darwin, 'Imperialism and the Victorians: The Dynamics of Territorial Expansion', *English Historical Review*, 112:447 (1997), pp. 614–42, on p. 629. For Darwin the concept of 'bridgeheads' was primarily introduced to 'help explain the baffling shape of the Victorian empire' and in particular the exchanges between formal and informal kinds of imperialism, but it also has potential when analysing dynamics and tensions within that 'baffling shape'.

16. A. Lester, 'Imperial Circuits and Networks: Geographies of the British Empire', *History Compass*, 4:1 (2006), pp. 124–41, on p. 129.

17. For a persuasive argument for the need to examine tensions among colonizers, see also A. L. Stoler and F. Cooper, 'Between Metropole and Colony: Rethinking a Research Agenda', in F. Cooper and A. L. Stoler (eds), *Tensions of Empire: Colonial Culture in a Bourgeois World* (Berkeley, CA: University of California Press, 1997), pp. 1–58.

18. Lester, 'Imperial Circuits and Networks', p. 129.

19. For an informed discussion of the differences in the theoretical assumptions of 'traditional' and 'new' imperial history, see D. Kennedy, 'Imperial History and Post-Colonial Theory', *Journal of Imperial and Commonwealth History*, 24:3 (1996), pp. 345–63. Any use of post-colonial theory in historical research does well to take seriously the thoughtful critique and reservations raised in D. Washbrooke, 'Orients and Occidents: Colonial Discourse Theory and the Historiography of British Imperialism', in Winks (ed.), *The Oxford History of the British Empire, Vol. V*, pp. 596–619.

20. For a good introduction and overview, see A. Picon, 'Engineers and Engineering History: Problems and Perspectives', *History and Technology*, 20:4 (2004), pp. 421–36. For classic essays set in the American context, see T. S. Reynolds (ed.), *The Engineer in America: A Historical Anthology from Technology and Culture* (Chicago, IL: Chicago University Press, 1991). For explorations in the European context, see K. Chatzis, 'Introduction: The National Identities of Engineers', *History and Technology*, 23:3 (2007), pp. 193–6, and the collection of articles in this special issue of that journal.

21. Buchanan, *The Engineers*, pp. 36–7.

22. R. A. Buchanan, '"Institutional Proliferation" in the British Engineering Profession 1847–1914', *Economic History Review*, n.s., 38:1 (1985), pp. 42–60.

23. Leading historians of engineering have only recently begun to analyse the category systematically and work has hitherto mainly concentrated on consulting engineers in the field of electricity. See S. Arapostathis, 'Consulting Engineers in the British Electric Light and Power Industry, c. 1880–1914' (unpublished PhD thesis, University of Oxford, 2006); S. Arapostathis, 'Morality, Locality and "Standardization" in the Work of the British Consulting Electrical Engineers, 1880–1914', in G. Gooday and J. Sumner (eds), *History of Technology*, 28, special issue: 'By Whose Standards? Standardization, Stability and Uniformity in the History of Information and Electrical Technologies' (2008), pp. 53–75; G. Gooday, 'Liars, Experts and Authorities', *History of Science*, 46:4 (2008), pp. 431–56.

24. W. V. Ball, *The Law Affecting Engineers* (London: A. Constable, 1909), p. 1.

25. The Association of Consulting Engineers was established in 1912 on the initiative of consultants in the bourgeoning field of electricity. For the consulting engineers in the field of civil engineering – and in particular for the Great George Street Clique – the institutional platform was the Institution of Civil Engineers. For the founding of the Association of Consulting Engineers, see H. Woodrow, *Tales of Victoria Street: The Story of the Association of Consulting Engineers* (London: Association of Consulting Engineers, 2003), pp. 6–18; Association of Consulting Engineers, *History of the Formation of the Association of Consulting Engineers* (London: Association of Consulting Engineers, 1930).

26. Ball, *The Law Affecting Engineers*, p. 5.

27. G. Watson, *The Civils: Story of the Institution of Civil Engineers* (London: Thomas Telford House, 1988), pp. 64–6.

28. R. S. Joby, *The Railway Builders: Lives and Works of the Victorian Railway Contractors* (London: David & Charles, 1983), pp. 145–7.

29. Woodrow, *Tales of Victoria Street*, pp. 3–6; Association of Consulting Engineers, *History of the Formation*, pp. 1–3.

30. D. H. Porter and G. C. Clifton, 'Patronage, Professional Values and Victorian Public Works: Engineering and Contracting the Thames Embankment', *Victorian Studies*, 31:3 (1988), pp. 321–49; Watson, *The Civils*, pp. 4–7.

31. Porter and Clifton 'Patronage', p. 334.

32. Darwin, 'Imperialism and the Victorians', pp. 640–1.

33. L. T. C. Rolt, *Victorian Engineering* (London: Allen Lane, 1970), p. 240.

34. The most important contributions to this reorientation include C. Macleod, *Heroes of Invention: Technology, Liberalism and British Identity, 1750–1914* (Cambridge: Cambridge University Press, 2007); B. Marsden and C. Smith, *Engineering Empires: A Cultural History of Technology in Nineteenth-Century Britain* (Hampshire: Palgrave Macmillan, 2007); G. Gooday, *Domesticating Electricity: Technology, Uncertainty and Gender, 1880–1914* (London: Pickering & Chatto, 2008); G. Gooday, 'Lies, Damned Lies and Declinism: Lyon Playfair, the Paris 1867 Exhibition and the Contested Rhetorics of Scientific Education and Industrial Performance', in I. Inkster (ed.), *The Golden Age: Essays in British Social and Economic History, 1850–1870* (Aldershot: Ashgate, 2000), pp. 105–21; D. Cannadine, 'Engineering History or the History of Engineering', *Transactions of the Newcomen Society*, 74B (2004), pp. 163–80.

35. The *locus classicus* for the argument of the decline of the industrial spirit is M. Weiner, *English Culture and the Decline of the Industrial Spirit, 1850–1980* (Cambridge: Cambridge University Press, 1982). For astute critiques of Weiner's argument of decline, see D. Edgerton, *Science, Technology and the British 'Decline' 1870–1970* (Cambridge: Cambridge University Press, 1996); and Macleod, *Heroes of Invention*, pp. 8–11.

36. Macleod, *Heroes of Invention*; Thompson, *The Empire Strikes Back?*, pp. 20–4.

37. R. A. Buchanan, 'Gentlemen Engineers: The Making of a Profession', *Victorian Studies*, 26:4 (1983), pp. 407–31; Marsden and Smith, *Engineering Empires*, pp. 254–8.

38. Marsden and Smith, *Engineering Empires*, pp. 239–41.

39. See D. Sunderland (ed.), *Communications in Africa, 1880–1939*, 5 vols (London: Pickering & Chatto, forthcoming 2012).

40. Until 1880 Crown Agents were also responsible for purchases and recruitment for railway projects in Cape Colony and Natal which were organized through the Westminster consulting engineer Charles Hutton Gregory. See A. J. Purkis, 'The Politics, Capital and

Labour of Railway-Building in the Cape Colony 1870–1885' (unpublished PhD thesis, University of Oxford, 1978), pp. 281–3.

41. For an informed introduction to theories and key concepts in the study of imperialism, see F. Cooper, *Colonialism in Question: Theory, Knowledge, History* (Berkeley, CA: University of California Press, 2005). For a concise history of the role of technology in Western imperialism, see D. R. Headrick, *Power over Peoples: Technology, Environments, and Western Imperialism, 1400 to the Present* (Princeton, NJ: Princeton University Press, 2010).

42. Buchanan, 'The Diaspora of British Engineering', p. 513.

43. J. Gallagher and R. E. Robinson, 'The Imperialism of Free Trade', *Economic History Review*, 2nd series, 6:1 (1953), pp. 1–15.

44. D. R. Headrick, *The Tentacles of Progress: Technology Transfer in the Age of Imperialism, 1850–1940* (Oxford: Oxford University Press, 1988), p. 379.

45. Naval Intelligence Department, *Handbook of Railways in Africa* (London: Admiralty, 1919), pp. 31–2.

1 Africa, Imperial Communication and the Engineering Press

1.   C. B. Going, 'The Engineer and the Policy of Imperialism', *Engineering Magazine – An Industrial Review*, 16:4 (1899), pp. 527–32, on p. 529.

2.   Marsden and Smith, *Engineering Empires*, p. 238.

3.   S. J. Potter, *News and the British World: The Emergence of an Imperial Press System* (Oxford: Oxford University Press, 2003), pp. 1–36.

4.   For an insightful historiographical perspective on producers, users and the construction of meanings in nineteenth-century periodicals, see J. Topham, 'Scientific Publishing and the Reading of Science in Nineteenth-Century Britain: A Historiographical Survey and Guide to Sources', *Studies in the History and Philosophy of Science*, 31:4 (2000), pp. 559–612. For important studies inspired by this approach, see G. Cantor, G. Dawson, G. Gooday, R. Noakes, S. Shuttleworth and J. Topham, *Reading the Magazine of Nature: Science in the Nineteenth-Century Periodical* (Cambridge: Cambridge University Press, 2004).

5.   J. Darwin, *After Tamerlane: The Global History of Empire since 1405* (London: Penguin, 2007), p. 298.

6.   Headrick, *The Tools of Empire*, pp. 9–13.

7.   For the ideological connections between technological ability and cultural superiority in Western imperialism, see M. Adas, *Machines as the Measure of Men: Science, Technology, and Ideologies of Western Dominance* (Ithaca, NY: Cornell University Press, 1989).

8.   F. Shelford, 'Sierra Leone in the Making', *Journal of the Royal African Society*, 28:111 (1929), pp. 235–40, on p. 238. For Shelford, see Chapters 3 and 5 of this volume.

9.   F. Fox, *River, Road, and Rail: Some Engineering Reminiscences* (London: John Murray, 1904), p. 192. For Fox, see Chapters 3 and 5 of this volume.

10. F. Fox, *Sixty-Three Years of Engineering, Scientific and Social Work* (London: John Murray, 1924), pp. 125–88.

11. G. L. Molesworth, *Imperialism and Free Trade* (London: E. & F. N. Spon, 1886); G. L. Molesworth, *Democracy and War* (London: E. & F. N. Spon, 1889); G. L. Molesworth, *Our Empire under Protection and Free Trade* (London: Lock & Co., 1902); G. L. Molesworth, *Economic and Fiscal Fallacies* (London: Longman, 1910).

12. Molesworth, *Our Empire under Protection*, p. 113.

13. J. Coode, 'Presidential Address', *PICE*, 99 (1889), pp. 1–40, on p. 39. For Coode, see Chapter 3 of this volume.
14. Marsden and Smith, *Engineering Empires*, pp. 236–41.
15. E. S. Ferguson, 'Technical Journals and the History of Technology', in S. H. Cutcliffe and R. C. Post (eds), *In Context: History and the History of Technology, Essays in Honour of Melvin Kranzberg* (New York: Lehigh University Press, 1989), pp. 53–71.
16. The useful metaphor of the marketplace in relation to historical analysis of periodical literature is developed and employed in A. Fyfe and B. Lightman (eds), *Science in the Marketplace: Nineteenth-Century Sights and Experiences* (Chicago, IL: University of Chicago Press, 2007).
17. Topham, 'Scientific Publishing'. For a discussion of the importance of publishers, see also P. J. Bowler, 'Experts and Publishers: Writing Popular Science in Early Twentieth-Century Britain', *British Journal for the History of Science*, 39:2 (2006), pp. 159–87. The *locus classicus* for the 'communication circle' in the history of the book is R. Darnton, *The Kiss of Lamourette: Reflections in Cultural History* (Markham: Penguin, 1990), pp. 107–35.
18. For a case study that argues for the need to take into consideration all sections of a publication to unravel its functions, see C. Andersen and H. H. Hjermitslev, 'Directing Public Interest: Danish Newspaper Science 1900–1903', *Centaurus: An International Journal for the History of Science*, 51:2 (2009), pp. 143–67.
19. *Engineering Magazine*, 16:2 (1899), p. 6.
20. After 1904 published as *Engineering Review*.
21. After 1914 published as *African and Eastern Engineering*.
22. The information on the history of *Engineer* in this paragraph is from B. Pendred (ed.), *The Engineer. Centenary Number: A Study of Influences of Engineering Advancement 1856–1956* (London: Morgan Brothers, 1956), pp. 146–9. The informative unsigned short history of the journal was in all likelihood written by Benjamin Pendred, the editor in 1946.
23. The information on the early history of *Engineering* in this paragraph is from Ferguson, 'Technical Journals', pp. 59–63. For a fascinating account of the life and career of Colburn, see J. Mortimer, *Zerah Colburn: Spirit of Darkness* (Bury St Edmunds: Arima, 2005).
24. The general information on composition, format and sales agencies are based on the two 'standard' issues of the journals: *Engineering* (31 May 1901), pp. 689–719; *Engineer*, 91 (31 May 1901), pp. 571–604.
25. This may well be due to the fact that advertisements in most instances have been cut when the journals were compiled in libraries. A collection with the advertisement sections fully intact is kept in Radcliffe Science Library, University of Oxford.
26. The first in the series of over a dozen articles was published in July 1902. See 'South Africa from an Engineer's Point of View', *Engineer*, 94 (18 July 1902), pp. 160–2.
27. 'South Africa and "the Engineer"', *Engineer*, 93 (6 June 1902), p. 563.
28. In revised form the article series was published in J. S. Ransome, *The Engineer in South Africa: A Review of the Industrial Situation in South Africa after the War and a Forecast of the Possibilities of the Country* (London: A. Constable, 1903).
29. 'South Africa from an Engineer's Point of View', *Engineer*, 94 (18 July 1902), pp. 160–1.
30. 'Engineers for Egypt', *Engineer*, 67 (27 June 1884), p. 485.
31. 'West African Railways', *Engineering*, 69 (13 April 1900), pp. 349–51, 362.
32. The coverage of the Aswan project in *Engineering* and *Engineer* is discussed in Chapter 6 of this volume.

33. 'The Uganda Railway I', *Engineer*, 94 (21 November 1902), pp. 491–2; 'The Uganda Railway II', *Engineer*, 94 (28 November 1902), pp. 512–13; 'The Uganda Railway III', *Engineer*, 94 (12 December 1902), pp. 570–2; 'The Uganda Railway IV', *Engineer*, 94 (26 December 1902), p. 612.

34. 'A Word of Introduction', *Engineering Magazine*, 14:1 (1897), p. 9.

35. *Engineering Magazine*, 15:3 (1898), p. 430. Going later became special lecturer in 'Industrial Engineering' at several universities in the United States, including Harvard and Columbia.

36. T. Feilden, 'No Apology', *Feilden's Magazine – The World's Record of Industrial Progress*, 1:1 (1899), p. 1.

37. 'Allen, Charles Edgar', in J. E. Sears (ed.), *Who's Who in Engineering 1920–21* (London: Compendium Publishing, 1921), p. 31.

38. C. B. Going, 'The Engineer and Imperialism', *Engineering Magazine*, 16:4 (1899), pp. 19–25, on p. 19.

39. Ibid., p. 19.

40. For an analysis of the symbolism in the frontispiece to *Novum Organon*, see P. Burke, *A Social History of Knowledge: From Gutenberg to Diderot* (Cambridge: Polity, 2000), pp. 113–15.

41. T. Carlyle, *Sartor Resartus* (1831; Oxford: Oxford University Press, 1999), p. 32.

42. Going, 'The Engineer and the Policy of Imperialism', p 527.

43. Ibid., p. 528.

44. Ibid., pp. 529–30.

45. J. Barrett, 'England, America, and Germany as Allies for the Open Door, I', *Engineering Magazine*, 17:6 (1899), pp. 893–902, on p. 895.

46. 'Review of the Continental Literature', *Engineering Magazine*, 17:6 (1899), p. 626.

47. H. G. Prout, 'The Economic Conquest of Africa' *Engineering Magazine*, 18:5 (1900), pp. 657–80; 'Prout H. G.', in *A & C's Who Was Who*, 3 vols (London: Adam & Charles Black, 1916–40), vol. 3: 1929–40, p. 1108.

48. Prout, 'The Economic Conquest of Africa', p. 665.

49. According to Marrison around the turn of the century there was 'a seemingly impossible number of policy objectives: Empire unity; imperial trade; retaliation and reciprocity with non-empire countries; protection at least to some degree'. See A Marrison, 'Comments on Howe and Sykes', in A. Marrison (ed.), *Free Trade and its Reception, 1815–1960* (London: Routledge, 1998), pp. 203–7, on pp. 204–5. For an analysis of the tariff reform debate in the imperial press system, see Potter, *News and the British World*, pp. 165–73.

50. C. B. Going, 'An Editorial Review: The Industrial Significance of the Anglo-German Alliance', *Engineering Magazine*, 20:3 (1901), pp. 325–31, on p. 329.

51. 'A Word of Introduction', *Engineering Magazine*, 14:1 (1897), p. 8.

52. J. R. Dunlap, 'Editorial Comment: Our Tenth Anniversary', *Engineering Magazine*, 21:1 (1901), p. 114.

53. Ibid.

54. N. Owen, 'Critics of Empire in Britain', in J. M Brown and W. R. Louis (eds), *The Oxford History of the British Empire, Vol. IV: The Twentieth Century* (Oxford: Oxford University Press, 1999), pp. 188–212, on p. 189.

55. *Feilden's Magazine*, 1:1 (1899), front page.

56. Feilden, 'No Apology', p. 1.

57. C. E. Allen, 'Lest we Forget', *Feilden's Magazine*, 1:1 (1899), pp. 4–5, on p. 4.

58. Ibid., p. 5.

59. For overviews and discussions, see S. Ledger and R. Luckhurst, *The Fin de Siècle: A Reader in Cultural History c. 1880–1900* (Oxford: Oxford University Press, 2000), pp. 133–73; B. Porter, *The Lion's Share: A Short History of British Imperialism*, 4th edn (London: Longman, 2004), pp. 203–5.

60. *Feilden's Magazine*, 3 (1901–2).

61. 'An Independent Platform', *Feilden's Magazine*, 2:3 (1900), p. 452.

62. *Feilden's Magazine*, 3:6 (1900), front page.

63. 'Interesting Announcements', *Feilden's Magazine*, 2:1 (1900), back cover.

64. 'A Letter from Brazil which Tells its own Story', *Engineering Magazine*, 22:2 (1901), p. 160.

65. 'Our Scope and Policy: Editorial Statement', *African Engineering*, 1:1 (1905), p. 9. Most likely the majority of these correspondents belonged to the network Leo Weinthal had established for *African World* (see note 67 below).

66. The information on Stafford Ransome's life and career is from 'Ransome, James Stafford', in W. H. Wills, *Anglo-African Who's Who and Biographical Sketch Book* (London: Upcott Gill, 1907), p. 248; 'Ransome, Stafford', in Sears (ed.), *Who's Who in Engineering*, pp. 198–9; 'Ransome, Stafford', in *A & C's Who Was Who*, vol. 3, p. 1123.

67. *African World – A Weekly Periodical of Africa*, 17:3 (1905). Leo Weinthal was a dedicated supporter of South African Union and edited a magnificent four-volume publication on the Cape–Cairo dream. See L. Weinthal (ed.), *The Story of the Cape to Cairo Railway and River Route, 1887–1922*, 4 vols in 5 (London: Pioneer Publishing Company, 1924).

68. 'Our New Departure', *African Engineering*, 3:44 (1908), p. 317.

69. 'African Engineering Notices', *African Engineering*, 1:10 (1905), p. 275.

70. *African Engineering*, 2:14 (1906), front page.

71. 'Editorial: Africa and the Engineer', *African Engineering*, 3:38 (1908), p. 121.

72. The existence of this department was first mentioned in 1905. See 'Business Information', *African Engineering*, 1:4 (1905), p. 71.

73. 'Business and Information Department', *African Engineering*, 1:6 (1905), p. 157.

74. 'Our Engineering Agency Department', *African Engineering*, 2:22 (1907), p. 407.

75. 'Our New Departure', *African Engineering*, 3:22 (1908), p. 317.

76. As will be discussed in Chapter 5, independently practising engineers were not allowed to advertise their services and it is likely that the business directory service also was a way of circumventing this ban.

77. *Commercial Pamphlet: African Engineering* (inserted in the binding of *African Engineering*, 3, in Rhodes House Archive, Oxford).

78. *African Engineering*, 3:27 (1907), front page.

79. Ransome was member of 'The Savage Club', frequented by many of London's artists and scientists. One book that benefited from Ransome's drawing talents was the political satire *Clara in Blunderland*, which he published with E. H. Begbie and M. H. Temple under the pseudonym Caroline Lewis. See C. Lewis, *Clara in Blunderland*, pictures by S. R. (London: William Heinemann, 1902).

80. 'Title and Allegory', *African Engineering*, 3:27 (1907), p. 11.

81. 'Our Scope and Policy: Editorial Statement', *African Engineering*, 1:1 (1905), p. 8.

82. 'Editorial Statement: All Parts of Africa', *African Engineering*, 4:40 (1908), front page.

83. 'The General Consultant', *African Engineering*, 5:51 (1909), p. 173; 'Conditions of Contract', *African Engineering*, 1:6 (1905), p. 73. See also Chapter 3 of this book.

84. For example in the 'Letters to the Editor' section and accompanying editorial, *African Engineering*, 1:3 (1905), p. 87.

85. 'Conditions of Contract', *African Engineering*, 1:6 (1905), p. 73.
86. 'Eastern Engineering', *African Engineering*, 5:66 (1910), p. 254.
87. For a critical analysis of the failure of the British Engineers Association and of Ransome's performance as secretary, see R. P. T. Davenport-Hines, 'The British Engineers' Association and Markets in China 1900–1930', in R. P. T. Davenport-Hines (ed.), *Markets and Bagmen: Studies in the History of Marketing and British Industrial Performance, 1830–1939* (London: Gower Publishing, 1987), pp. 102–30.
88. 'Some Changes but No Change of Policy', *African Engineering*, 7:90 (1912), p. 261; 'Dunell, Herbert', in Sears (ed.), *Who's Who in Engineering*, p. 93.
89. 'African Engineering and Eastern Engineering', *African Engineering*, 9:112 (1914), p. 96.
90. Darwin, *After Tamerlane*, p. 303.
91. The perceptive phrase was formulated by Leonard Reich and is quoted in Ferguson, 'Technical Journals', p. 70.
92. After Ransome resigned from the British Engineers Association in 1916 he entered the family business which by then had expanded and relocated from London to Newark-upon-Trent. See 'Ransome, A & Co Ltd', in Sears (ed.), *Who's Who in Engineering*, p. 346.

2 Engineers in Imperial London

1. Quoted from H. Lefebvre, 'The Other Parises', in *Henri Lefebvre: Key Writings*, ed. S. Elden, E. Lebas and E. Kofman (London: Continuum, 2003), pp. 151–8, on p. 156.
2. D. Keene, 'Cities and Empires', *Journal of Urban History*, 32:1 (2005), pp. 8–21 provides a good overview and introduction to this theme.
3. The metaphor of an imperial map of London is introduced in F. Driver and D. Gilbert, 'Imperial Cities: Overlapping Territories, Intertwined Histories', in Driver and Gilbert (eds), *Imperial Cities*, pp. 2–17.
4. D. Lambert and A. Lester (eds), *Colonial Lives Across the British Empire: Imperial Careering in the Long Nineteenth Century* (Cambridge: Cambridge University Press, 2005).
5. For introductions to 'the spatial turn' across fields and disciplines, see B. Warf and S. Aris, *The Spatial Turn: Interdisciplinary Perspectives* (London: Routledge, 2008); D. Bachmann-Medick, *Cultural Turns, Neuorientierungen in den Kulturwissenschaften* (Reinbek: Rowohlt, 2006).
6. E. W. Soja, *Postmodern Geographies – The Reassertion of Space in Critical Social Theory* (London: Verso Books, 1989).
7. D. N. Livingstone, *Putting Science in its Place – Geographies of Scientific Knowledge* (Chicago, IL: Chicago University Press, 2003), pp. 7–8.
8. H. K. Bhaba, *The Location of Culture* (London: Routledge, 1994); E. Said, *Culture and Imperialism* (New York: Vintage Books, 1994), pp. 1–72.
9. For hybridity of place, see J. N. Pieterse, 'Globalization as Hybridization', in M. Featherstone, S. Lash and R. Robertson (eds), *Global Modernities* (London: Sage Publications, 1995), pp. 45–67; J. R. Short, *Urban Theory: A Critical Assessment* (Hampshire: Palgrave Macmillan, 2006).
10. Driver and Gilbert, 'Imperial Cities', p. 3.
11. D. Massey, *Space, Place and Gender* (Minneapolis, MN: University of Minnesota Press, 1994), pp. 149–53.
12. J. Schneer, *London 1900: The Imperial Metropolis* (London: Yale University Press, 1999), pp. 13, 37–63.

13. R. Michie, *The City of London: Continuity and Change since 1850* (Hampshire: Palgrave Macmillan, 1992).
14. P. J. Cain and A. G. Hopkins, *British Imperialism 1688–2000* (London: Longman, 1993), pp. 121–32, quote on p. 122.
15. M. H. Port, *Imperial London: Civil Government Building in London 1851–1914* (London: Yale University Press, 1995), p. 14. For maps of Westminster and a broad introduction to the history of the area since Roman times, see S. Bradley and N. Pevsner, *London 6: Westminster* (London: Yale University Press, 2003).
16. Port, *Imperial London*, pp. 5–25.
17. Schneer, *London 1900*, p. 17.
18. The number of engineers registered with business addresses in London at the time. The numbers are based on *Kelly's Directories of the Engineers, Iron and Metal Trades 1883* (London: Kelly & Co., 1883), 'Civil Engineers', pp. 26–30; *Kelly's Directories of the Engineers, Iron and Metal Trades 1890* (London: Kelly & Co., 1890), 'Civil Engineers', pp. 627–31; *Kelly's Directories of the Engineers, Iron and Metal Trades 1901* (London: Kelly & Co., 1901), 'Civil Engineers', pp. 794–8; *Kelly's Directories of the Engineers, Iron and Metal Trades 1909* (London: Kelly & Co., 1909), 'Civil Engineers', pp. 854–6.
19. The number of engineers registered in the south-west postal district. The SW postcode was established in 1867 and it covered Westminster, St James's and Mayfield, the central areas of London's West End. Changes in the postal code boundary were not made until 1917 when the district was divided into sections numbered 1–10. The postal museum and archive provides detailed information on the history of postal districts in London. In this case see in particular http://www.postalheritage.org.uk/history/downloads/ BPMA_Info_Sheet_Postcodes_web.pdf [accessed 25 April 2011].
20. The number of engineers with business addresses in 'Central Westminster'; the streets immediately bordering the broad avenue of Whitehall. For the year 1883 the streets included in this category are Whitehall Place, Parliament Street, Great George Street, Victoria Street (including Westminster Chambers and Victoria Chambers), Delahay Street, Queen Anne's Gate and Queen Anne Place. In the period under study the central area of Westminster underwent large-scale restructuring where the street layouts and terraces changed considerably. By 1890 Storey Gate and the Sanctuary had been rebuilt and housed several engineering offices which belong to the category of 'Central Westminster'. Victoria Chambers and Westminster Chambers had been demolished but Victoria Street together with Great George Street remained the place where most London civil engineers resided. From the turn of the century the north side of Great George Street was torn down to make space for a host of new government offices. The street thereafter housed few private businesses. Most of the engineers' offices that remained in the area of 'Central Westminster' moved to Victoria Street and to the rebuilt Tothill Street between Victoria Street and St James's Park. Issues arising from the changing cityscape are impossible to avoid as renaming and renumbering in the period 1889–1912 affected about 5,000 streets and terraces in the county of London, about one quarter of the total. For an assessment of this problem, see P. J. Atkins, *The Directories of London 1677–1977* (London: Continuum, 1990), pp. 120–33.
21. Ibid., pp. 131–2.
22. This aspect and the general functions of the council of the ICE are analysed in detail in Chapter 4. The point that the most prestigious segment of the engineering profession clustered in Westminster is also supported by the addresses of the presidents of the ICE during this period; see Table 3.1.

23. Location of council members are from addresses stated in the published membership lists of the ICE; Institution of Civil Engineers, *Charter, Supplemental Charter, By-Laws and Lists of Members 1885* (London: Institution of Civil Engineers, 1885), p. 57; Institution of Civil Engineers, *Charter, Supplemental Charter, By-Laws and Lists of Members 1895* (London: Institution of Civil Engineers, 1895), p. 57; Institution of Civil Engineers, *Charter, Supplemental Charter, By-Laws and Lists of Members 1902* (London: Institution of Civil Engineers, 1902), p. 57; Institution of Civil Engineers, *Charter, Supplemental Charter, By-Laws and Lists of Members 1912* (London: Institution of Civil Engineers, 1912), p. 57. Prior to 1896 past presidents of the Institution were automatically listed as council members and the number of members therefore varied. A reform in 1896 changed his procedure so that only four past presidents could be members of council. After this the number of council members was each year 35.

24. Studies of locations from the perspective of economic geography go back to A. Marshall, *The Economics of Industry* (London: John Murray, 1888) and have exploded in new directions in particular since 1990. Classic contributions that address the question of clustering include M. E. Porter, *The Competitive Advantages of Nations* (San Francisco, CA: Jossey Bass, 1990); and P. Krugman, *Geography and Trade* (Boston, MA: MIT Press, 1991). A. Saxenian, *Regional Advantage: Culture and Competition in Silicon Valley and Route 128* (Cambridge, MA: Harvard University Press, 1994) studies the location often used as the paradigmatic example of modern clustering.

25. M. Ball and D. Sunderland, *An Economic History of London 1800–1914* (London: Routledge, 2000).

26. Ibid., pp. 332–3 (for accountants), pp. 166–7 (for publishers), pp. 329–31 (for legal professions).

27. Port, *Imperial London*, p. 165.

28. See C. Dickens, *Little Dorrit* (1857; New York: Digireads, 2009), pp. 72–85, for the elegant political satire 'Containing the Whole Science of Government'.

29. M. Chrimes, 'John Fowler: Engineer or Manager?', *PICE*, 97 (1993), pp. 135–43, makes this point in relation to the career of John Fowler.

30. T. C. Fidler, *Civil Engineering* (London: Methuen & Co., 1905), p. 84; For Fidler's Westminster address, see *Kellys Directories 1901*, p. 795.

31. Fidler, *Civil Engineering*, p. 90.

32. The list is based on *Kelly's Mayfair, St. James's, Soho and Westminster Directory 'Buff Book' 1887* (London: Kelly & Co., 1887), pp. 67–72 (government offices), pp. 216–18 (colonial representations). The office of the Crown Agents in Millbank, however, does not appear in the directory and for this information I am indebted to the publishers' anonymous manuscript reviewer.

33. 'Great Engineering Societies of Europe', *Engineering Magazine*, 26:4 (1904), p. 520.

34. Ball and Sunderland, *An Economic History of London*, pp. 21–2.

35. Information on the location of these contractors is from *Kelly's 'Buff Book' 1887*, p. 124 (Lucas & Aird), p. 221 (Ransome & Rapier), p. 178 (Pauling & Co.).

36. Ball and Sunderland, *An Economic History of London*, pp. 315–16.

37. M. E. W. Williams, *The Precision Makers: A History of the Instruments Industry in Britain and France 1870–1939* (London: Thomson Learning, 1994), pp. 14–17, 43–6.

38. Fidler, *Civil Engineering*, pp. 91–2.

39. Rough estimates of numbers of mechanical and electrical engineers are based on *Kelly's Directories* for the years 1890 and 1909.

40. Ball and Sunderland, *An Economic History of London*, p. 21.

41. Potter, *News and the British World*.
42. For a useful analysis of the functions of gentlemen's clubs in London in this period, see A. Taddei, 'The London Club in Late Nineteenth Century', *Discussion Paper in Economic and Social History April 1999*, at http://econpapers.repec.org/paper/nufesohwp/_5F028 [accessed 29 September 2009].
43. F. R. Cowell, *The Athenaeum: Club and Social Life in London 1824–1974* (Oxford: Heinemann, 1975).
44. *Kelly's 'Buff Book' 1887*, pp. 77–9.
45. Information on the club membership of the individual engineers is based on their entries in *A & C's Who Was Who*, vols 1–3. The orientation of the clubs is based on the designations used in *Kelly's 'Buff Book' 1887*, pp. 77–9.
46. R. Barton, '"Huxley, Lubbock, and Half a Dozen Others": Professionals and Gentlemen in the Formation of the X Club 1851–1864', *Isis*, 89:3 (1998), pp. 410–44, on p. 429.
47. For a full list and a short history of the most important British engineering associations, see Sears (ed.), *Who's Who in Engineering*, pp. 5–21.
48. Ibid., p. 9.
49. Ibid., p. 11.
50. Ibid., p. 10.
51. Ibid., p. 14.
52. Ball and Sunderland, *An Economic History of London*, p. 124.
53. R. Appleyard, *The History of the Institution of Electrical Engineers 1871–1931* (London: Institution of Electrical Engineers, 1939), pp. 21–3, 159–60, 186–93.
54. Buchanan, '"Institutional Proliferation"', pp. 47–8.
55. R. H. Parsons, *A History of the Institution of Mechanical Engineers 1847–1947* (London: Institution of Mechanical Engineers, 1947), pp. 21–2, 36–9.
56. Ibid., p. 38.
57. Institution of Civil Engineers, *A Brief History of the Institution of Civil Engineers with an Account of the Charter Centenary Celebration June 1928* (London: Institution of Civil Engineers, 1928), pp. 17–19.
58. For a full analysis of the restructuring of Westminster, see Port, *Imperial London*, pp. 114–38.
59. 'Report of Special Housing Committee', 14 March 1892, CMB, 13, p. 403.
60. 'Draft Petition', 14 March 1892, CMB, 13, p. 405.
61. Port, *Imperial London*, p. 54.
62. M. Dunkeld, 'Paraphernalia of the Professions – The Case of the Institution of Civil Engineers' (unpublished manuscript), p. 9. I wish to thank Malcolm Dunkeld for sharing his wide knowledge of the ICE building and for allowing me to quote from this excellent manuscript, which has yet to be published.
63. Institution of Civil Engineers, *Report of the Proceedings at the Ceremony of Laying the Foundation Stone of the New Building on Tuesday, the 25th October, 1910* (London: Institution of Civil Engineers), p. 11.
64. 'The New Home of the Civil Engineers', *African Engineering*, 6:65 (1910), p. 323.
65. Dunkeld, 'Paraphernalia of the Professions', pp. 61–2.
66. Ibid., p. 62.
67. The most comprehensive history of this controversial railway project is M. F. Hill, *The Permanent Way: The Story of the Kenya and Uganda Railway* (Nairobi: East African Railways, 1957). Other histories include R. Hardy, *The Iron Snake* (Glasgow: Collins, 1965); and C. Miller, *The Lunatic Express – An Entertainment in Imperialism* (London:

Penguin, 1977). The latter can be appreciated for its literary qualities rather than scholarly level. See also H. Gunston, 'The Planning and Construction of the Uganda Railway', *Journal of Newcomen Society*, 74A (2004), pp. 45–71.

68. 'George Whitehouse, the Constructor of the Uganda Railway', *Graphic* (20 June 1903), pp. 702–3; Hill, *The Permanent Way*, p. 145.

69. S. J. North, *Europeans in British Administrated East Africa: A Biographical Listing* (London: S. J. North, 2005), p. 509.

70. George Whitehouse, Diary entries 9 September–8 November 1897, Rhodes House Archive, Oxford, MSS Afr. S.1046 (3); Whitehouse, Diary entries 7 May–8 November 1900, MSS Afr. S. 1046 (6).

71. Whitehouse, Diary entries 18–21 June 1900, MSS Afr. S. 1046 (6).

72. For the political build-up, see J. S. Galbraith, *Mackinnon and East Africa, 1875–1895: A Study in the 'New Imperialism'* (Cambridge: Cambridge University Press, 1972).

73. Hill, *The Permanent Way*, p. 132.

74. North, *Europeans in British Administrated East Africa*, p. xv.

75. I. J. Kerr, 'Sir Francis Langford O'Callaghan', in *ODNB*, at http://oxforddnb.com/view/article/35282 [accessed 11 April 2011].

76. Hill, *The Permanent Way*, pp. 201–3.

77. Whitehouse, Diary entry 24 July 1899, MSS Afr. S. 1046 (5); Hill, *The Permanent Way*, p. 191.

78. Whitehouse, Diary entries 10 March–16 April 1900, MSS Afr. S. 1046 (6).

79. Hill, *The Permanent Way*, p. 196.

80. Whitehouse, Diary entries 3 July, 13 July, 21 August, 6 September 1900, MSS Afr. S. 1046 (6).

81. T. R. Metcalf, *Imperial Connections: India in the Indian Ocean Arena 1860–1920* (Los Angeles, CA: University of California Press, 2007), pp. 165–204.

82. S. Morris, 'Indians in East Africa: A Study in a Plural Society', *British Journal of Sociology*, 7: 3 (1956), pp. 194–211.

83. Hazareesingh, 'Interconnected Synchronicities', p. 8.

84. Whitehouse, Diary entry 12 June 1903, MSS Afr. S. 1046 (9).

85. North, *Europeans in British Administrated East Africa*, p. xv.

86. Whitehouse was also discontented that his replacement in 1903 was offered a starting salary of £1,500. See Whitehouse, Diary entry 17 June 1903, MSS Afr. S. 1046 (9).

87. O'Callaghan to Whitehouse, 23 April 1896, George Whitehouse Correspondence 1896–1902, MSS Afr. S. 1046 (11), p. 1 (coal supplies); Whitehouse to O'Callaghan, 17 May 1897, Whitehouse Correspondence 1896–1902, MSS Afr. S. 1046 (11), p. 12 (terms of leave).

88. Whitehouse, Diary entries March–June 1901, MSS Afr. S. 1046 (7).

89. F. Jackson, *Early Days in East Africa* (London: Collins & Co, 1931), p. 321.

90. This is revealed in Whitehouse's final dispatch to the Railway Committee, handed in just after the severance of his connection with the railway 1 April 1903. See Whitehouse to the Chairmen and Members of Uganda Railway Committee, 13 May 1903, Whitehouse Correspondence 1896–1902, MSS Afr. S. 1046 (11), p. 180.

91. Whitehouse, Diary entry 31 April 1903, MSS Afr. S. 1046 (9). Whitehouse was replaced as chief engineer by the assistant chief engineer on the railway since 1895, Frank Rawson. By then only minor works were left to be completed.

92. Whitehouse, Diary entry 13 May 1903, MSS Afr. S. 1046 (9).

93. Whitehouse, Diary entries 13 August, 23 August 1903, MSS Afr. S. 1046 (9).

94. Whitehouse, Diary entry 14 May 1903, MSS Afr. S. 1046 (9).
95. Whitehouse, Diary entry 23 September 1903, MSS Afr. S. 1046 (9).
96. 'Obituary: Sir George Whitehouse', *The Times*, 21 July 1938.
97. 'Obituary: Charles O. Burge', *PICE*, 134 (1912), pp. 213–14.
98. C. O. Burge, *The Adventures of a Civil Engineer: Fifty Years on Five Continents* (London: Alston Rivers, 1909), p. 50.
99. Ibid., p. 136.
100. Ibid., p. 130.
101. Ibid., p. 217.
102. C. O. Burge, 'The Hawkesbury Bridge, New South Wales', *PICE*, 101 (1890), pp. 2–12.
103. Burge, *The Adventures of a Civil Engineer*, pp. 304–20.
104. D. Lambert and A. Lester, 'Introduction: Imperial Spaces, Imperial Subjects', in Lambert and Lester (eds), *Colonial Lives*, pp. 1–31, on pp. 1–6.

3 Engineering Networks and the Great George Street Clique

1. H. Perkin, *The Rise of Professional Society: England since 1880* (London: Routledge, 1989), p. 4.
2. These were mainly landowners and businessmen but also included a few exceptionally successful lawyers, artists, authors, publishers, surgeons and engineers. For an introduction to the elite stratum in the period 1880–1914, see ibid., pp. 61–80. For engineers specifically, see Buchanan, 'Gentlemen Engineers'.
3. For a solid overview of the historiography on networks and British imperialism, see Magee and Thompson, *Empire and Globalisation*, pp. 22–32.
4. J. Scott, *Social Network Analysis. A Handbook: Theory and Analysis* (1991; London: Sage Publications, 2000).
5. Laidlaw, *Colonial Connections*, p. 13.
6. D. Sunderland, *Social Capital, Trust and the Industrial Revolution 1780–1880* (London: Routledge, 2008), pp. 50–67, 85–95.
7. R. I. Rotberg (ed.), *Patterns of Social Capital: Stability and Change in Historical Perspective* (Cambridge: Cambridge University Press, 2000); F. Fukuyama, *Trust: The Social Virtues and the Creation of Prosperity* (London: Penguin, 1996); R. D. Putnam, *Bowling Alone: The Collapse and Revival of America Community* (New York: Simon & Schuster, 2001).
8. J. F. Wilson and A. Popp (eds), *Industrial Clusters and Regional Business Networks in England 1750–1970* (Oxford: Oxford University Press, 2003).
9. Sunderland, *Social Capital*, pp. 6–11.
10. M. Biagioli, *Galileo Courtier: The Practice of Science in the Culture of Absolutism* (Chicago, IL: Chicago University Press, 1993). For a discussion of networks and patronage in science and technology in the *long durée*, see B. Mazlich, 'Technology and Social Relations: From Patronage to Networks', in P. Lyth and H. Trischler (eds), *Wiring Prometheus: Globalisation, History and Technology* (Aarhus: Aarhus University Press, 2004), pp. 21–35.
11. For patronage and Renaissance engineers, see Misa, *Leonardo to the Internet*, pp. 1–33.
12. The rise of middle-class professions is well covered in the literature. See, for example, Perkin, *The Rise of Professional Society*. For engineers, see W. J. Reader, *Professional Men: The Rise of Professional Classes in Nineteenth-Century England* (London: Weidenfeld & Nicholson, 1966), pp. 118–36; Buchanan, *The Engineers*, pp. 161–80.
13. Porter and Clifton, 'Patronage'; Watson, *The Civils*, pp. 6–8.

14. Laidlaw, *Colonial Connections*, pp. 14–15.
15. Porter and Clifton, 'Patronage', p. 334.
16. M. Chrimes, 'John Fowler', in *ODNB*, at http://oxforddnb.com/view/article/99655 [accessed 20 March 2011]; 'Obituary: Sir John Fowler', *PICE*, 135 (1898), pp. 328–37.
17. T. Mackay, *The Life of Sir John Fowler – Engineer* (London: John Murray, 1900), pp. 59–60, 118–19.
18. Chrimes, 'John Fowler: Engineer or Manager?'.
19. R. Lee, *The Greatest Public Work: The New South Wales Railways, 1848–1889* (Sydney: Southwood Press, 1988), pp. 49–52. I wish to thank Robert Lee for sharing generously of his unsurpassed knowledge of railway construction in Australia.
20. R. Lee, *Colonial Engineer: John Whitton 1819–1898 and the Building of Australia's Railways* (Sydney: University of New South Wales Press, 2000), pp. 308–9. The relationship between Fowler and Whitton is analysed in detail in Lee's fine biography of Whitton and is placed in a wider imperial context in C. Andersen, 'Colonial Connections and Consulting Engineers 1850–1914', *Engineering History and Heritage*, 165:1 (forthcoming 2011).
21. Mackay, *The Life of Sir John Fowler*, pp. 188–9.
22. Command Papers; Accounts and Papers LXIII.611/C9331, 'Report on the Uganda Railway by Sir Guilford Lindsey Molesworth', pp. 1–60, on p 12, at http://gateway.proquest.com/openurl?url_ver=Z39.88-2004&res_dat=xri:hcpp&rft_dat=xri:hcpp:rec:1899-077292 [accessed 26 April 2011].
23. For Ismail's schemes to modernize Egypt, see F. R. Hunter, *Egypt under the Khedives, 1805–1879: From Household Government to Modern Bureaucracy* (Cairo: American University Press in Cairo, 1999).
24. Mackay, *The Life of Sir John Fowler*, pp. 229–30.
25. J. Fowler, 'Egyptian Reports as the General Engineering Advisor to the Egyptian Government 1869–77', Archive of the ICE, no. 62, M 603A.
26. J. Fowler, Report of Darfour Railway and on the Soudan Railway Extension to Khartoum', in ibid., pp. 2–3.
27. Mackay, *The Life of Sir John Fowler*, p. 249.
28. Cain and Hopkins, *British Imperialism*, pp. 312–14.
29. A. L. A. Sayyid-Marsot, 'The British Occupation of Egypt from 1882', in Porter (ed.), *The Oxford History of the British Empire, Vol. III*, pp. 651–64.
30. F. W. Spear (rev. M. Chrimes), 'Benjamin Baker', in *ODNB*, at http://oxforddnb.com/view/article/30545 [accessed 20 March 2011]; 'Obituary: Sir Benjamin Baker', *PICE*, 170 (1907), pp. 377–84.
31. B. Baker, 'Cleopatra's Needle', *PICE*, 62 (1878), pp. 233–44.
32. A full list of Baker's publications in *PICE* was given in his obituary in that journal and included 'The River Nile', 'Cleopatra's Needle, 'The Practical Strength of Beams', 'The Actual Lateral Pressure of Earthwork', 'Steel for Tires and Axles' and 'The Metropolitan and Metropolitan District Railways'.
33. H. G. C. Ketchum, *The Chignecto Ship Railway* (Boston, MA: Damrell & Upham, 1893).
34. 'Obituary: Sir Benjamin Baker', pp. 382–4.
35. 'Special Council Meeting', 18 February 1900, CMB, 15, p. 313.
36. A. E. Shelford, *The Life of Sir William Shelford, KCMG, Chevalier of the Order of the Crown of Italy, Member of Council of the Institution of Civil Engineers* (privately printed, 1909), p. 14; 'Obituary: Sir William Shelford', *PICE*, 163 (1905), pp. 384–6.
37. Shelford, *The Life of Sir William Shelford*, pp. 19, 36–48.

38. Ibid., pp. 101–13.
39. 'Obituary: Sir William Shelford', p. 384.
40. Shelford, *The Life of Sir William Shelford*, pp. 113–18, 121–34.
41. The expression 'old comrade' was used in Benjamin Baker's obituary in *PICE* to describe the relationship between the William Shelford and Baker. See 'Obituary: Sir Benjamin Baker', p. 382.
42. 'The Death of Capt. F. Shelford', *Journal of the Royal African Society*, 42:169 (1943), p. 182.
43. For extensive archival material on Coode including the work of Coode & Son in Ceylon and Africa, see Coode Archive: Reports and Correspondence Selected from the Records of Coode & Partners, Archive of the ICE, no. 627.2M.
44. T. H Beare (rev. A M. Wood), 'John Coode', in *ODNB*, at http://oxforddnb.com/view/article/10022 [accessed 20 March 2011].
45. 'Obituary: John Coode', *PICE*, 113 (1893), pp. 134–43.
46. 'Obituary: Arthur Trevenen Coode', *Journal of the Institution of Civil Engineers*, 16:5 (1941), pp. 80–1.
47. For a detailed analysis of the founding and first years of the BSAC, see J. S. Galbraith, *Crown and Charter – The Early Years of British South Africa Company* (Berkeley, CA: University of California Press, 1974).
48. R. Thorne, 'Charles Fox', in *ODNB*, at http://oxforddnb.com/view/article/10022 [accessed 20 March 2011]; 'Obituary: Charles Fox', *PICE*, 75 (1875), pp. 264–6.
49. Fox, *Sixty-Three Years of Engineering*, pp. 21–2.
50. R. Freeman, '(Charles) Douglas Fox', in *ODNB*, at http://oxforddnb.com/view/article/37428 [accessed 20 March 2011).
51. A. Thomas, *Rhodes: The Race for Africa* (London: St Martin's Press, 1996), p. 236.
52. 'Obituary: Sir Charles Douglas Fox', *PICE*, 213 (1922), pp. 414–18.
53. For a detailed but dated history of the establishment of the Rhodesian railway system, see A. H. Croxton, *Railways of Rhodesia: The Story of the Beira, Mashonaland and Rhodesian Railways* (Devon: David & Charles, 1974). For a pungent critique of the historiographical assumptions that guide the tradition of railway history to which Croxton's book belongs, see I. R. Phimister, 'Towards a History of Zimbabwe's Railways', *Zimbabwean History*, 12 (1981), pp. 61–91.
54. E. I. Carlyle (rev. S. Katzenellenbogen), 'Charles Herbert Theophilus Metcalfe', in *ODNB*, at http://oxforddnb.com/view/article/35001 [accessed 20 March 2011]; 'Obituary; Sir Charles Herbert Theophilus Metcalfe', *PICE*, 228 (1928), pp. 350–2.
55. Reader, *Professional Men*, pp. 117–19.
56. Shelford, *The Life of Sir William Shelford*, pp. 19–20.
57. Sunderland, *Social Capital*, pp. 136–44.
58. Reader, *Professional Men*, pp. 119–21; Porter and Clifton, 'Patronage', pp. 331–7.
59. 'Obituary: Charles Beresford Fox', *PICE*, 189 (1912), p. 348; Fox, *River, Road and Rail*, pp. 173–7.
60. M. R. Lane, *The Rendel Connection: A Dynasty of Engineers* (London: Quiller Press, 1989), pp. 71–2; Whitehouse, Diary entry 14 January 1903, MSS Afr. S. 1046 (9).
61. Shelford, *The Life of Sir William Shelford*, pp. 111, 116–17.
62. 'Southern Nigeria Lagos Harbour Survey 1910', Coode Archive, ICE, no. 627.2M.
63. Buchanan, *The Engineers*, p. 187.
64. F. Shelford, 'Pioneer Engineering I', *Engineer*, 104 (8 May 1908), pp. 469–71, on p. 469.

65. 'The information in Table 3.1 is based on obituaries of the presidents published in the *PICE*, except Guildford Molesworth, which is based on E. J. Molesworth, *Life of Sir Guildford Molesworth: The Nestor of the Engineering Profession* (London: E. & F. N. Spon, 1922), and Robert Elliott-Cooper, which is based on 'Obituary: Robert Elliott-Cooper', *Journal of the Institution of Civil Engineers*, 18:6 (1942), pp. 229–30. John Fowler and John Hawkshaw were presidents prior to 1880 but because their influence continued into the period they are included in the table.

66. 'Obituary: John Hawkshaw', *PICE*, 106 (1891), pp. 321–35.

67. M. Chrimes, 'British Civil Engineers in Canada 1830–90', *Proceedings of CSCE International Symposium on Civil Engineering History*, 1 (2005), pp. 3–28, on p. 17.

68. 'Obituary: Harrison Hayter', *PICE*, 134 (1898), pp. 391–4; 'Obituary: John C. Hawkshaw', *PICE*, 215 (1923), p. 321.

69. 'Obituary: John Wolfe Barry', *PICE*, 206 (1918), pp. 350–7.

70. Mackay, *The Life of Sir John Fowler*, pp. 213–15.

71. Ibid., p. 236. In 1900 Mackay had access to Fowler's personal papers and based his description of Fowler's experiences in Egypt on these. Unfortunately, the papers have been lost since.

72. Ibid., pp. 267–75.

73. Few rulers are ever fully independent; Egypt was nominally under the rule of the Ottoman empire and the sovereignty of the khedival regime was further constrained by the European powers in particular after the opening of the Suez Canal, see Hunter, *Egypt under the Khedives*, pp. 179–227.

74. R. L. Tignor, 'The "Indianisation" of the Egypt Administration under British Rule', *American Historical Review*, 68:3 (1963), pp. 636–61.

75. R. H. Brown, *History of the Barrage at the Head of the Delta of Egypt*, introductory note W. E. Garstin (Cairo: Public Works Print, 1896), pp. 20–1.

76. For a detailed description of the repair works on the barrages, see R. L. Tignor, 'British Agricultural and Hydraulic Policy in Egypt 1882–1892', *Agricultural History*, 37:2 (1963), pp. 63–74, on pp. 65–9.

77. The agreement between Cromer, Cassel and Aird is analysed in R. Owen, *Lord Cromer – Victorian Imperialist, Edwardian Proconsul* (Oxford: Oxford University Press 2004), pp. 292–4.

78. Gifford to Colonial Office, 3 January 1889, Colonial Office Papers 372/245/3 [372: Confidential print, Africa South]. Lord Gifford was the driving force in the Exploring Company. For an analysis of his relations with the Colonial Office, see P. Maylam, *Rhodes, the Tswana, and the British: Colonialism, Collaboration and Conflict in the Bechuanaland Protectorate 1885–1899* (London: Greenwood Press, 1980), pp. 49–59.

79. P. Maylam, 'The Making of the Kimberley–Bulawayo Railway', *Rhodesian History*, 8 (1977), pp. 13–35.

80. C. Metcalfe, 'My Story of the Scheme', in Weinthal (ed.), *Story of the Cape to Cairo Railway*, vol. 1, pp. 91–9.

81. R. I. Rotberg, *The Founder – Cecil Rhodes and the Pursuit of Power* (London: Southern Book Publishers, 1988), pp. 406–8; Metcalfe, 'My Story of the Scheme', vol. 1, p. 97.

82. 'Royal Charter', 29 October 1889, Colonial Office Papers 372/245/163.

83. Galbraith, *Crown and Charter*, provides the fullest account the formative years of the BSAC.

84. R. Phimister, 'Rhodes, Rhodesia and the Rand', *Journal of Southern African Studies*, 1:1 (1974), pp. 74–91.

85. J. Lunn, 'The Political Economy of Primary Railway Construction in Rhodesia', *Journal of African History*, 33:2 (1992), pp. 239–54.

86. The apt notion of 'the charmed circle' of the BSAC is employed by Lunn in ibid., p. 245.
87. Croxton, *Railways of Rhodesia*, p. 63; Rotberg, *The Founder*, p. 46.
88. In two books David Sunderland has supplied a critical and superb analysis of the diverse and changing functions of the Crown Agents. See D. Sunderland, *Managing the British Empire: The Crown Agents, 1833–1914* (Woodbridge: Boydell Press, 2004); D. Sunderland, *Managing British Colonial and Post-Colonial Development: The Crown Agents 1914–74* (Woodbridge: Boydell Press, 2007).
89. Shelford, *The Life of Sir William Shelford*, pp. 133–5. For a discussion of Chamberlain's programme of 'constructive imperialism', see Porter, *The Lion's Share*, pp. 198–202.
90. Sunderland, *Managing the British Empire*, p. 61.
91. Frederic Shelford Papers, 'Press Cuttings', Rhodes House Archive, Oxford, MSS Afr. S. 2193, box 1.
92. D. Sunderland, 'Montagu Ommanney', in *ODNB*, at http://oxforddnb.com/view/article/57565 [accessed 20 March 2011].
93. Sunderland, *Managing the Empire*, pp. 57–62, 110–16.
94. D. Sunderland, 'The Departmental System of Railway Construction in British West Africa, 1895–1906', *Journal of Transport History*, 23:2 (2002), pp. 87–112.
95. A good impression of Frederic Shelford's high societal life can be obtained from the collections of press cuttings in the Shelford collection in Rhodes House Archive.
96. See Table 3.1 and Andersen, 'Colonial Connections'. For a full list of presidents of the ICE, see Watson, *The Civils*, pp. 251–5.
97. E. C. Baker, *Preece and Those who Followed: Consulting Engineers in the Twentieth Century* (Dublin: Reprographic, 1980), pp. 3–6, 34–43.
98. E. C. Baker, *Sir William Preece, Victorian Engineer Extraordinary* (London: Hutchinson & Co., 1976), pp. 310–12; 'Sivewright, Sir James', in *Dictionary of South African Biography*, 5 vols (Pretoria: Nasionale Boekhandel; Durban: Butterworth & Co., 1968–87), vol. 4, pp. 572–4.
99. Baker, *Preece and Those who Followed*, pp. 177–81.
100. Macleod, *Heroes of Invention*, pp. 236–45.
101. 'Letter to the Editor: The Crown Agents as Purchasers', *African Engineering*, 1:2 (1905), p. 17.
102. 'Letter to the Editor: The Consultant and the Manufacturer', *African Engineering*, 1:5 (1905), p. 87.
103. 'Leader: The General Consultant', *African Engineering*, 5:51 (1909), pp. 173–4.
104. J. D. Hargreaves, *West Africa Partitioned*, 2 vols (London: Macmillan, 1985), vol. 2.
105. Sunderland, *Managing the Empire*, pp. 143–4.
106. Ibid., pp. 145–6.
107. For a study of Girouard's career as engineer, soldier and colonial administrator, see A. Kirk-Greene, 'Canada in Africa: Sir Percy Girouard, Neglected Colonial Governor', *African Affairs*, 83:331 (1984), pp. 207–39.
108. F. Jaekel, *History of the Nigerian Railway*, 3 vols (Ibadan: Spectrum Books, 1997), vol. 2, pp. 98–101.
109. Percy Girouard to Frederick Lugard, 25 January 1908, Lugard Papers, Rhodes House Archive, Oxford, MSS Afr. R. 231, MSS Brit Emp. S. 63, f. 191.
110. Girouard to Lugard, 25 January 1908, in Lugard Papers, MSS Afr. R. 231, MSS Brit Emp. S. 63, ff. 193–4.
111. Jaekel, *History of the Nigerian Railway*, vol. 2, pp. 94–5.

112. Girouard to Lugard, 25 January 1908, Lugard Papers, MSS Afr. R. 231, MSS Brit Emp. S. 63, f. 195.

113. A. R. Seymour, 'Tropical Railway', Rhodes House Archive, Oxford, MSS Afr. R. 213, ff 40.

114. Ibid., ff. 32–42.

115. Ibid., f. 49.

116. For photos and a description of station buildings designed by Shelford & Son for West African railways, see 'Talks with Engineers: Mr. Shelford on West African Railway Development', *African Engineering*, 1:6 (1905), pp. 135–9, on p 138; 'The Sierra Leone Railway', *Engineer*, 87 (8 May 1899), p. 346.

117. Sunderland, *Managing the Empire*, p. 145.

118. Lugard to Undersecretary of State for the Colonies, 10 May 1912, Lugard Papers, MSS Afr. R. 231, MSS Brit Emp. S. 74, f. 190.

119. Sunderland, *Managing the Empire*, p. 146.

120. F. Shelford, *Address on West African Railways, the African Trade Section of the Incorporated Chamber of Commerce Liverpool* (Liverpool: Liverpool Chamber of Commerce, 1900), p. 3.

121. Girouard to Lugard, 25 January 1908, Lugard Papers, MSS Afr. R. 231, MSS Brit Emp. S. 63, f. 201.

4 Empire in the Institution of Civil Engineers

1. B. Baker, 'Presidential Address', *PICE*, 123 (1896), pp. 1–38, on p. 36.

2. Buchanan, '"Institutional Proliferation"'.

3. For a listing of membership numbers in the largest British engineering institutions, see Buchanan, *The Engineers*, p. 233.

4. Two other membership categories existed at the time. Honorary members were elected among scientists, politicians and other men of notability and influence. A class of associates consisted of members who were not practising engineers but still were engaged in the 'science of engineering'. The best overview of membership categories and election criteria during this period is provided in Fidler, *Civil Engineering*, pp. 169–79. For a fuller description and discussion of changing membership criteria in the ICE, see Watson, *The Civils*, pp. 50–2, 111–30.

5. Watson, *The Civils*, pp. 83–94.

6. For a list of degrees from universities in Britain and colonies approved by the ICE, see Fidler, *Civil Engineering*, pp. 175–7.

7. Watson, *The Civils*, pp. 117–18.

8. B. H. Beckett, *Scientific London* (London: John Murray, 1875), pp. 76–136.

9. Watson, *The Civils*, pp. 143–6.

10. Porter and Clifton, 'Patronage', p. 338.

11. Marsden and Smith, *Engineering Empires*, p. 237.

12. A notable exception is Chrimes, 'British Civil Engineers in Canada', which focuses on ICE members in Canada but also draws informative comparisons with developments across the British empire up to 1890.

13. Institution of Civil Engineers, *A Brief History*, p. 18.

14. Watson, *The Civils*, pp. 14, 55–6.

15. Numbers and percentages are based on 'President's Memorandum I: Constitution and Election of the Council', CMB, 14, p. 346.

16. For an informed discussion of the work of ICE members in Argentina in this period, see M. Chrimes, 'British and Irish Civil Engineers in the Development of Argentina in the Nineteenth Century', in M. Dunkeld et al. (eds), *Proceedings of the Second International Congress on Construction History*, 1 (2006), pp. 675–94.

17. Institution of Civil Engineers, *A Brief History*, pp. 74–7.

18. Numbers in the table are calculated from Institution of Civil Engineers, *Charter 1885*; Institution of Civil Engineers, *Charter 1902*; Institution of Civil Engineers, *Charter 1912*.

19. Excluding students and honorary members.

20. Numbers in the table are calculated from, Institution of Civil Engineers, *Charter 1885*; Institution of Civil Engineers, *Charter 1902*; Institution of Civil Engineers, *Charter 1912*.

21. Excluding students and honorary members.

22. Watson, *The Civils*, pp. 54–5.

23. 'Special Meeting', 17 December 1895, CMB, 14, pp. 342–50.

24. 'President's Memorandum I: Constitution and Election of the Council', CMB, 14, pp. 346, 344, 343.

25. 'Special Meeting', 17 December 1895, CMB, 14, p. 342.

26. The first supplemental charter was introduced in 1887 in connection with the erection of a new building in Great George Street, see Institution of Civil Engineers, *Charter, Supplemental Charter, By-Laws and Lists of Members 1887* (London: Institution of Civil Engineers, 1887), pp. 11–13; and Watson, *The Civils*, p. 58. Apart from the enlargement of the council and the possibility of placing members based in the colonies for election to the council, the 1896 reform also introduced examinations of applicants for associate membership.

27. Institution of Civil Engineers, *Charter, Supplemental Charter, By-Laws and Lists of Members 1897* (London: Institution of Civil Engineers, 1897), pp. 17–20.

28. Ibid., p. 57.

29. 'Council Meeting', 1 March 1898, CMB, 15, p. 45.

30. Institution of Civil Engineers, *Charter, Supplemental Charter, By-Laws and Lists of Members 1898* (London: Institution of Civil Engineers, 1898), p. 57.

31. Baker, 'Presidential Address', p. 36.

32. 'Report of the Council, Session 1897–98', *PICE*, 134 (1898), pp. 208–9.

33. J. Wolfe Barry, 'Presidential Address', *PICE*, 134 (1898), pp. 1–13, on p. 3.

34. 'Council Meeting', 4 March 1902', CMB, 16, p. 263.

35. Institution of Civil Engineers, *Charter 1902*, p. 57.

36. 'Council Meeting', 19 March 1902, CMB, 16, p. 66. On this occasion Thomas Stewart of the Cape Colony was allowed re-nomination for reasons that were not further specified in the minutes of meeting. This appears to have been the only departure from the two-year rule.

37. T. Murray, 'Thomas Stewart – First South African Consulting Engineer', *Documents of the Association of Consulting Engineers*, at http://email.asce.org/international/documents/05.MurrayStewartTWoodhead.pdf [accessed 30 November 2010].

38. 'Obituary, John Brown', *PICE*, 106 (1906), pp. 403–4.

39. 'Report of the Council', 1 March 1910, CMB, 19, p. 426.

40. 'The Institution of Civil Engineers in Cape Town', *Engineer*, 94 (26 September 1902), p. 306.

41. Ibid.

42. 'Memorandum on Overseas Advisory Committees', in Archive of the ICE, Overseas Advisory Committees – General History from 1880, no. 187/185. This memorandum on the history of advisory committees was drawn up in 1929 when the council contemplated reforming its relations with the committees. The collection in Overseas Advisory Committees – General History from 1880 contains comprehensive archival material relating to the establishment of advisory committees across the empire. The analysis presented here, however, focuses primarily on the committees in southern Africa.

43. 'Council Meeting', 30 November 1890, CMB, 11, p. 320.

44. It is difficult to establish exact dates for the establishment of the advisory committees, as communication between the council and members in the colonies was slow. The foundation years stated here are based on the years when the council was contacted by colonial members regarding the founding of a committee in the specific colony. Advisory committees were established in Canada in 1911 and in India in 1912. See 'Memorandum on Overseas Advisory Committees', p. 8.

45. For an early discussion of this among council members, see 'Council Meeting', 29 March 1892, CMB, 12, p. 335.

46. 'Council Meeting', 1 February 1898, CMB, 15, pp. 20–4.

47. 'Council Meeting', 1 February 1898, CMB, 15, pp. 21–2.

48. 'Council Meeting', 7 February 1893, CMG, 13, p. 252.

49. For material relating to the Transvaal committee, see Transvaal Correspondence.

50. S. H. Farrar to Council, 2 June 1894, Transvaal Correspondence.

51. S. H. Farrar to Council, 9 June 1894, Transvaal Correspondence.

52. S. H. Farrar to Council, 2 June 1894, Transvaal Correspondence.

53. S. H. Farrar to J. Forest, 20 October 1894, Transvaal Correspondence.

54. S. H. Farrar to Council, 23 September 1896, Transvaal Correspondence.

55. Institution of Civil Engineers, *Charter, Supplemental Charter, By-Laws and Lists of Members 1896* (London: Institution of Civil Engineers, 1896), p. 187.

56. There are several accounts of the Jameson Raid, see among others C. Saunders and I. R. Smith, 'Southern Africa, 1795–1910', in Porter (ed.), *The Oxford History of the British Empire, Vol. III*, pp. 597–624; and Thomas, *Rhodes*, pp. 273–304.

57. 'Farrar George Herbert', in *Dictionary of South African Biography*, vol. 1, pp. 287–8; Thomas, *Rhodes*, p. 304.

58. S. H. Farrar to Council, 29 September 1896, Transvaal Correspondence.

59. E. B. J. Knox to Council, 19 December 1896, Transvaal Correspondence.

60. S. H. Farrar to Council, 14 December 1896, Transvaal Correspondence.

61. 'List of Members and Balloting List', December 1896, Transvaal Correspondence.

62. For material relating to the South African committee, see Advisory Committee for South Africa.

63. 'Council Meeting', 22 December 1903, CMB, 17, p. 101; 'Note on the Constitution of Advisory Committee in South Africa', Advisory Committee for South Africa.

64. 'Council Meeting', 22 January 1907, CMB, 18, p. 121.

65. 'Council Meeting', 17 January 1911, CMB, 20, p. 22.

66. A. M. Tippet to Tudsbury, 12 March 1914, Advisory Committee for South Africa.

67. A. M. Tippet to Tudsbury, 12 March 1914, Advisory Committee for South Africa.

68. Tudsbury to A. M. Tippet, 29 April 1914, Advisory Committee for South Africa.

69. Stewart to Tudsbury, 25 May 1914, Advisory Committee for South Africa.

70. Nicolson to Council, 9 April 1927, Advisory Committee for South Africa.

71. Institution of Civil Engineers, *Charter 1912*, pp. 60–1.

72. According to Watson the first advisory committee was set up in Buenos Aires in 1927 but this was, however, a reinvention of overseas advisory committees. For a discussion of the development of advisory committees from 1927, see Watson, *The Civils*, pp. 234–8.

73. ICE to Engineer-in-Chief, Whitehouse Correspondence 1896–1902, MSS Afr. S. 1046 (12), p. 179.

74. Institution of Civil Engineers, *A Brief History*, p. 40. Here it was rightly asserted that 'should some future historian turn his attention to the subject of engineering in the 19th and 20th century these pamphlets may well prove a mine of curious information'.

75. Watson, *The Civils*, pp. 140–1.

76. F. Shelford, 'Some Features of the West African Government Railways. Including Appendix', *PICE*, 189 (1912), pp. 1–23.

77. 'Discussion: Some Features of the West African Government Railways', *PICE*, 189 (1912), pp. 22–45, on p. 25.

78. 'Correspondence: Some Features of the West African Government Railways', *PICE*, 189 (1912), pp. 45–80.

79. Watson, *The Civils*, p. 142.

80. J. A. Johnson, 'Germany: Discipline – Industry – Profession: German Chemical Organisations', in A. K. Nielsen S. Štrbáňová, *Creating Networks in Chemistry: The Founding and Early History of Chemical Societies in Europe* (London: RSC Publishing, 2008), pp. 113–38, on p. 133.

81. Watson, *The Civils*, p. 147.

82. 'Council Meeting', 16 February 1896, CMB, 14, p. 576.

83. 'Engineering Conference 25–27 May 1897', *PICE*, 130 (1897), p. 201; 'Engineering Conference', *PICE*, 138 (1899), p. 369.

84. 'The Engineering Conference', *Engineering* (14 June 1907), p. 779.

85. 'Engineering Conference 25, 26, 27 May 1897', *PICE*, 130 (1897), pp. 174–6, on p. 175.

86. Ibid., p. 176.

87. 'Council Meeting', 5 February 1907, CMB, 18, p. 249.

88. 'The Conference Excursions: The Riverside London Water Companies', *Engineer*, 83 (4 June 1897), p. 564.

89. 'Institution of Civil Engineers: Engineering Conference', *Engineering* (1 May 1903), p. 595.

90. Marsden and Smith, *Engineering Empires*, pp. 229–31.

91. 'The Conversazione', *Engineering* (28 June 1907), p. 673.

92. 'The Conversazione of the Institute [*sic*] of Civil Engineers', *Engineer*, 83 (28 May 1897), p. 550.

93. See J. M. Mackenzie, *Propaganda and Empire: Manipulation of British Public Opinion, 1880–1960* (Manchester: Manchester University Press, 1986), pp. 96–121. For a more comprehensive analysis, see P. H. Hoffenberg, *An Empire on Display: English Indian and Australian Exhibitions from Crystal Palace to the Great War* (Berkeley, CA: University of California Press, 2001).

94. W. Beinart and L. Hughes, *Environment and Empire* (Oxford: Oxford University Press, 2007), p. 225.

95. For a detailed discussion of the representation of machines and technology in imperial exhibitions, see Hoffenberg, *An Empire on Display*, pp. 166–222.

96. 'Engineering Conference Programme', *Engineering* (14 June 1907), pp. 630–1.

97. 'Review of the British Press', *Engineering Magazine*, 18:5 (1899), p. 838.

98. 'The Engineering Conference', *Engineering* (19 June 1903), p. 817.

99. 'President's Memorandum I: Constitution and Election of the Council', CMB, 14, p. 346.

100. 'Report of the Council', *PICE*, 130 (1896), p. 102.

101. 'Engineering Conference: Address, John Wolfe Barry', *PICE*, 130 (1896), p. 175.

102. 'Conference of Civil Engineers', *The Times*, 17 May 1897.

103. 'Leader: The Conference', *Engineer*, 83 (21 May 1897), p. 515.

104. 'Engineering Conference: Address, William H. Preece', *PICE*, 138 (1899), pp. 363–4.

105. Ibid., p. 364.

106. W. H. Preece, 'The Engineering Conference 1899', *Feilden's Magazine* 1:1 (1899), pp. 2–15 on p. 13.

107. The format of the conferences was altered slightly after 1907. In 1911 – and in accordance with the four-year interval between the conferences that was becoming the norm – the gathering at the ICE was devoted to a specific theme: 'Training and Education of Engineers'. In 1915 no conference was held due to the outbreak of the Great War. When the tradition of general conferences was re-invoked after the war the engineers dealt with in this book had left the stage.

108. Buchanan, 'The Diaspora of British Engineering'. For the many pitfalls in relying on this source for quantitative studies of engineers, see C. Macleod and A. Nuvolari, 'The Pitfalls of Prosopography: Inventors in the Dictionary of National Biography', *Technology and Culture*, 47:4 (2006), pp. 757–76.

109. Laidlaw, *Colonial Connections*, pp. 14–15.

5 Explorer-Engineers and Gentlemen in the Public Eye

1. P. Høeg, 'Journey into a Dark Heart', in P. Høeg, *Tales of the Night* (1990; London: Penguin, 1999), pp. 3–38, on p. 8. I wish to thank Henry Nielsen for making this fine short story, set at the inauguration of a Cabinda–Katanga railway in 1929, known to me.

2. K. Chew and A. Wilson, *Victorian Science and Engineering portrayed in The Illustrated London News* (London: Allan Sutton, 1993).

3. S. Smiles *The Collected Works of Samuel Smiles* (1857–1905; London: Routledge, 1997).

4. For a superb analysis of the characteristics of 'Smilesian heroes', see G. Cantor, 'The Scientist as Hero: Public Images of Michael Faraday', in M. Shortland and R. Yeo (eds), *Telling Lives in Science: Essays on Scientific Biography* (Cambridge: Cambridge University Press, 1996), pp. 171–93.

5. A. Jarvis, *Samuel Smiles and the Construction of Victorian Values* (Gloucestershire: Sutton Publishing, 1997).

6. Weiner, *English Culture*.

7. Macleod, *Heroes of Invention*; Cannadine, 'Engineering History'; Gooday, 'Lies, Damned Lies and Declinism'.

8. Macleod, *Heroes of Invention*, p. 378.

9. B. Riffenburgh, *The Myth of the Explorer: The Press, Sensationalism, and Geographical Discovery* (Oxford: Oxford University Press, 1994); J. M. MacKenzie, 'The Iconography of the Exemplary Life: The Case of David Livingstone', in G. Cubitt and A. Warren (eds), *Heroic Reputations and Exemplary Lives* (Manchester: Manchester University Press, 2000), pp. 84–104.

10. For an informed discussion of the 'afterlife' of the era of the great explorers, see F. Driver, *Geography Militant: Cultures of Exploration and Empire* (Oxford: Blackwell Publishers, 2001), pp. 199–219.

11. A. G. Hopkins, 'Explorers' Tales: Stanley Presumes – Again', *Journal of Imperial and Commonwealth History*, 36:4 (2008), pp. 669–84; Driver, *Geography Militant*, pp. 117–46.

12. H. M. Stanley, *The Congo and the Founding of its Free State; a Story of Work and Exploration* (London: Sampson Low, 1885), pp. 182–4.

13. W. Churchill, *My African Journey* (London: Hodder & Stoughton, 1908), p. 51. Churchill's description of his African journey with the Uganda Railway was first published as two articles in the popular *Strand Magazine*.

14. F. A. Talbot, *The Railway Conquest of the World* (Philadelphia, PA: J. B. Lippincott, 1911), pp. 2, 9–10; F. A. Talbot, *Railway Wonders of the World* (London: Cassell, 1917).

15. J. Hawkshaw, *Some Reminiscences of South America: From Two and a Half Years' Residence in Venezuela* (London: Jackson & Walford, 1838).

16. Burge, *The Adventures of a Civil Engineer*, p. 86.

17. See, among others, M. L. Pratt, *Imperial Eyes: Travel Writing and Transculturation* (London: Routledge, 1992). For an informed discussion of the virtues and limitations of such 'textual' approaches to the study of exploration, see D. Kennedy, 'British Exploration in the Nineteenth Century: A Historiographical Survey', *History Compass*, 5:6 (2007), pp. 1879–900.

18. For an extended survey of this familiar yet crucial theme, see Adas, *Machines as the Measures of Men*.

19. P. Girouard, 'Autobiography', Rhodes House Archive, Oxford, MSS Afr. S. 1865, ff. 16–17. The manuscript is not dated, but references to ongoing events indicate that it was written in the early 1920s. For a biographical study of Girouard's career, see Kirk-Greene, 'Canada in Africa'. For Girouard's writing for a lay audience, see, among others, P. Girouard, 'The Railways of Africa', *Scribner's Magazine – Monthly with Illustrations*, 39 (1906), pp. 553–69.

20. F. Shelford, 'Pioneer Engineering III', *Engineer*, 104 (22 May 1908), pp. 528–9. Shelford explicitly used the term native carrier as 'a generic name for all unskilled labour'.

21. See, for example, D. Brooke, *The Railway Navvy: 'That Despicable Race of Men'* (Devon: Davis & Charles, 1983).

22. A. Crozier, 'Sensationalising Africa: British Medical Impressions of Sub-Saharan Africa 1890–1939', *Journal of Imperial and Commonwealth History*, 35:3 (2007), pp. 393–415; For a parallel perspective which introduces the term 'tropicality', see F. Driver and L. Martins (eds), *Tropical Visions in an Age of Empire* (Chicago, IL: Chicago University Press, 2005).

23. The bridge is also known as the Victoria Falls Bridge. In referring to it as the Zambezi Bridge I follow Douglas Fox & Partners' chief designer G. A. Hobson. See G. A. Hobson, 'The Great Zambezi Bridge', in Weinthal (ed.), *The Story of the Cape to Cairo Railway*, vol. 1, pp. 43–59.

24. Phimister, 'Towards a History of Zimbabwe's Railways', pp. 82–3; S. Katzenellenbogen, *Railways and the Copper Mines of Katanga* (Oxford: Oxford University Press, 1973), pp. 46–9.

25. 'The Victoria Falls Bridge', *Engineer*, 104 (28 January 1905), pp. 339–41.

26. See, among others, L. Mitchell, 'The Cape to Cairo Railway', *Journal of the Society of Arts*, 55:2822 (1906), pp. 98–109; Weinthal (ed.), *Story of the Cape to Cairo Railway*. For an excellent analysis of the Cape–Cairo imagery, see P. Merrington, 'A Staggered Orientalism: The Cape to Cairo Imagery', *Poetics Today*, 22:2 (2001), pp. 323–64.

27. The chief designer of the bridge later confirmed this legend but also noted that: 'Apart from all sentimental considerations, the site is admirable. Indeed from personal examination, I am of the opinion that it is the best possible position for a bridge', see Hobson, 'The Great Zambezi Bridge', vol. 1, p. 45.

28. See J. R. Ryan, *Picturing Empire: Photography and the Visualization of the British Empire* (Chicago, IL: Chicago University Press, 1998).

29. N. Starostina, 'Engineering the Empire of Images: Constructing Railways in Asia before the Great War', *Southeast Review of Asian Studies*, 31 (2009), pp. 181–206.

30. See, among others, 'The Cape to Cairo Railway No. II – Bulawayo to Khartoum', *Engineer*, 102 (14 August 1904), p. 160; Hobson, 'The Great Zambezi Bridge', vol. 1, pp. 48–9, 52.

31. Fox, *River, Road, and Rail*, p. 179.

32. For a collection of Zambezi Bridge postcards and memorabilia, see 'A Study of the Victoria Falls Bridge using Old Postcards and Photographs' at http://www.geoffs-trains.com/Bridge/bridgehome.html [accessed 23 April 2011].

33. For the visit of British Association for the Advancement of Science to South Africa see, S. Dubow 'A Commonwealth of Science: The British Association in South Africa, 1905 and 1929', in S. Dubow (ed.), *Science and Society in Southern Africa* (Manchester: Manchester University Press, 2000), pp. 66–100. Dubow notes that engineers were very active in the formative years of the South African Association for the Advancement of Science.

34. For Metcalfe's role in organizing the visit, see Croxton, *Railways of Rhodesia*, pp. 76–7.

35. 'The British Association in South Africa', *The Times*, 13 September 1905.

36. Croxton, *Railways of Rhodesia*, pp. 190–2.

37. This section draws on C. Andersen, 'Explorer-Engineers take the Field: Imperial Engineers, Africa and the Late-Victorian Public', in M. Harbsmeier, K. H. Nielsen and C. J. Ries (eds), *Scientists and Scholars in the Field: Studies in the History of Expeditions and Fieldwork* (Aarhus: Aarhus University Press, forthcoming 2011).

38. J. H. Patterson, *The Man-Eaters of Tsavo and Other East African Adventures* (1907; New York: Hard Press, 2003). Besides Patterson's own account the story of the Tsavo lions is treated in detail in Miller, *The Lunatic Express*, pp. 242–99. Hardy, *The Iron Snake*, pp. 141–70, also tells the story and portrays Patterson as a violent, racist tyrant.

39. Patterson, *The Man-Eaters of Tsavo*, p. xii.

40. Rhodes House Archive, Oxford, Papers and Diaries of John Henry Patterson, 'Press Cuttings', MSS Afr. R. 93.

41. For more balanced assessment of the impact of Tsavo incident than what is offered by Miller, Hardy and Patterson, see Hill, *The Permanent Way*, pp. 131–42. Hill's assessment is also supported by Patterson private account of the events as they unfolded. See J. H. Patterson, Diary entries February 1898–January 1899, MSS Afr. R. 93.

42. For an informative though not scholarly biography of Patterson, see D. Brian, *The Seven Lives of Colonel Patterson: How an Irish Lion Hunter Led the Jewish Legion to Victory* (New York: Syracuse, 2008).

43. Patterson, Diary entry 4 March 1898, MSS Afr. R. 93.

44. Brian, *The Seven Lives of Colonel Patterson*.

45. Whitehouse, Diary entries 13 May 1903–18 May 1903, MSS Afr. S. 1046 (9).

46. Whitehouse, Diary entry 18 May 1903, MSS Afr. S. 1046 (9). On this day he was interviewed by Mr Emmett of Reuter. The interview was printed in a number of national

and regional papers For a full collection, see Whitehouse Correspondence 1896–1903, 'Collections of Press Cuttings', MSS Afr. S. 1046 (11).

47. North, *Europeans in British Administrated East Africa*, p. xv.

48. 'George Whitehouse, the Constructor of the Uganda Railway', *Graphic* (20 June 1903), in Whitehouse Correspondence 1896–1903, 'Collections of Press Cuttings', MSS Afr. S. 1046 (11).

49. 'The Chief-Engineer of the Uganda Railway', *Evening Post*, 21 May 1903, in Whitehouse Correspondence 1896–1903, 'Collections of Press Cuttings', MSS Afr. S. 1046 (11).

50. B. Whitehouse, 'To the Victoria Nyanza by the Uganda Railway', *Journal of the Society of Arts*, 38:2 (1902), pp. 229–41.

51. For Benjamin Whitehouse's drawings and reports of the survey, see B. Whitehouse, 'Account of Survey of the Northern Section of Lake Victoria', Rhodes House Archive, Oxford, MSS Afr. S. 1294.

52. Whitehouse, 'To the Victoria Nyanza by the Uganda Railway', p. 238.

53. Ibid., p. 241.

54. 'The Society of Arts', *Morning Post*, 29 January 1902, in Whitehouse Correspondence 1896–1903, 'Collections of Press Cuttings', MSS Afr. S 1046 (11).

55. 'The Uganda Railway', *The Times*, 29 January 1902, in Whitehouse Correspondence 1896–1903, 'Collections of Press Cuttings', MSS Afr. S 1046 (11).

56. G. T. Goldie, 'Letter to the Editor: The Uganda Railway', *The Times*, 20 January 1902. Whitehouse, Diary entry, 11 December 1901, MSS Afr. S. 1046 (7).

57. 'Cartoon: The Uganda Railway', *Liverpool Daily Chronicle*, 6 March 1903, in Whitehouse Correspondence 1896–1903, 'Collections of Press Cuttings', MSS Afr. S 1046 (11).

58. See also Sunderland, 'The Departmental System of Railway Construction'.

59. 'A New Route to Kumassi', *Graphic* (1 July 1899), pp. 7–8, in Shelford Papers, 'Press Cuttings', MSS Afr. S. 2193, box 1.

60. Other publications that contain photos and drawings of Shelford in West Africa include F. Shelford, 'To Kumassi by Rail', *Graphic* (22 December 1900), pp. 935–6; F. Shelford, 'Survey of the Niger', *Graphic* (24 June 1907), pp. 1154–5; 'The Sierra Leone Railway', *Engineer*, 87 (8 September 1899), pp. 241–3; Talbot, *Railway Wonders of the World*, pp. 61–72.

61. 'Talks with Engineers: Mr. Shelford on West African Railway Development', *African Engineering*, 1:6 (1905), pp. 135–9, on p. 135.

62. Ibid., pp. 138–9. Henry Seton Merriman was the pseudonym used by Hugh Stowell Scott when he wrote his West African story *With Edged Tools*.

63. 'Though the Jungle', *Financier* (22 January 1901), in Shelford Papers, 'Press Cuttings', MSS Afr. S. 2193, box 1.

64. Shelford Papers, 'Press Cuttings', MSS Afr. S. 2193, box 1.

65. Shelford, *Address on West African Railways*; 'Railways in West Africa [Report from banquet hosted by Royal Colonial Institute]', *The Times* 13 April 1904; 'Sekondi to Kumasi. Progress of the Railway. Mr Shelford gives the latest progress report [Banquet to Colonel James Willcocks at the Hotel Metropole]', *Financier* (9 March 1901), in Shelford Papers, 'Press Cuttings', MSS Afr. S. 2193, box 1.

66. J. D. Fage, 'When the African Society was Founded: Who were the Africanists?', *African Affairs*, 94:376 (1995), pp. 369–81; Royal African Society, 'Fifty Years of a British African Society', *African Affairs*, 50:200 (1951), pp. 177–95; D. J. Birkett, *Mary Kingsley: Imperial Adventuress* (Hampshire: Palgrave Macmillan, 1992).

67. 'The Study of African Questions', *St James's Gazette*, 28 June 1901, in Shelford Papers, 'Press Cuttings', MSS Afr. S. 2193, box 1; Shelford, *The Life of Sir William Shelford*, pp. 138–40.

68. F. Shelford, 'Some Notes on the History of the African Society', *Journal of the Royal African Society*, 34:136 (1935), pp. 223–6, on p. 224.

69. F. Shelford, 'The Late Mrs J. R. Green and the African Society', *Journal of the Royal African Society*, 28:112 (1929), p. 414.

70. For Chamberlain and 'constructive imperialism', see Porter, *The Lion's Share*, pp. 198–202.

71. Shelford, 'Sierra Leone in the Making'; F. Shelford, 'Ten Years of Progress in West Africa', *Journal of the Royal African Society*, 6:24 (1907), pp. 341–9.

72. F. Shelford, 'On West African Railways', *Journal of the Royal African Society*, 1:3 (1902), pp. 339–54, on p. 340.

73. Shelford, *Address on West African Railways*, p. 5.

74. Shelford, 'Pioneer Engineering I'; F. Shelford, 'Pioneer Engineering II', *Engineer*, 104 (15 May 1908), pp. 495–6; Shelford, 'Pioneer Engineering III'; F. Shelford, 'Pioneer Engineering IV', *Engineer*, 104 (29, May 1908), pp. 549–50.

75. F. Shelford, *Pioneering* (London: E. & F. N. Spon, 1909).

76. Shelford, 'Pioneer Engineering I', p. 469.

77. Shelford, 'Pioneer Engineering II', p. 496.

78. Institution of Civil Engineers, *Charter 1895*, p. 191; Institution of Civil Engineers, *Charter 1902*, p. 111.

79. F. Shelford, 'Railway Surveying in Tropical Forest', *PICE*, 133 (1898), pp. 339–50.

80. 'Obituary: Sir Robert Elliott-Cooper', *Journal of the Institution of Civil Engineers*, 18:6 (1942), pp. 229–30.

81. House of Commons Papers; Accounts and Papers LVI. 637/103, 'The Remuneration of the Consulting Engineers to Crown Colonies and Protectorates', at http://gateway.proquest.com/openurl?url_ver=Z39.88-2004&res_dat=xri:hcpp&rft_dat=xri:hcpp:rec:1907-007668 [accessed 26 April 2011].

82. 'Obituary: Frederic Shelford', *The Times*, 2 August 1943.

83. 'The Death of Captain F. Shelford', *Journal of Royal African Society*, 42:169 (1943), p. 182.

84. C. Macleod, 'The Nineteenth-Century Engineer as Cultural Hero', in A. Kelly and M. Kelly (eds), *Brunel: In Love with the Impossible* (Reston, VA: American Society of Civil Engineers, 2006), pp. 61–79; Macleod, *Heroes of Invention*, pp. 2–6; Marsden and Smith, *Engineering Empires*, pp. 227–30.

85. Marsden and Smith, *Engineering Empires*, p. 227.

86. R. A. Buchanan, *Brunel: The Life and Times of Isambard Kingdom Brunel* (London: Continuum, 2002), pp. 25–30.

87. 'The Design of Rolling Stock for Rhodesian Railways', *Engineering* (31 May 1901), pp. 694–5.

88. 'The Kumasi Railway', *Engineering* (31 May 1901), pp. 689–90, 693.

89. 'The Assouan Dam', *Engineer*, 67 (11 August 1899), pp. 140, 142–3; 'The Assouan Masonry Dam on the Nile', *Engineering* (9 March 1900), pp. 315–16, 318.

90. 'The Assouan Dam', *Engineer*, p. 142.

91. J. Fowler and B. Baker, 'A Sweet-Water Ship Canal through Egypt', *Nineteenth Century*, 13:1 (1883), pp. 166–72.

92. Ibid., p. 172.

93. C. Metcalfe, and F. I. Richarde-Seaver, 'The British Sphere of Influence in South Africa', *Fortnightly Review*, 267 (1889), pp. 351–63. The article's co-author was a mining investor with stakes in the BSAC. For Richarde-Seaver, see Rotberg, *The Founder*, p. 233.

94. For the public charter campaign of the BSAC, see Thomas, *Rhodes*, pp. 206–11. This aspect is somewhat neglected in Galbraith's in other respects superior study of the founding of the BSAC.

95. Metcalfe, 'My Story of the Scheme', p. 233.

96. Metcalfe and Richarde-Seaver, 'The British Sphere of Influence', p. 361.

97. See, among others, C. Metcalfe *The Cape-to-Cairo Line: How the War Might Solve a Problem* (London: Royal Empire Society, 1914); C. Metcalfe, 'Railway Development of Africa, Present and Future', *Geographical Journal*, 47:1 (1916), pp. 3–17; Metcalfe, 'My Story of the Scheme'. Metcalfe was also frequently interviewed on the progress of the Cape–Cairo Railway for *The Times* and other British papers.

98. Metcalfe, 'Railway Development of Africa', p. 14.

99. Ibid., p. 17.

100. 'Obituary: Sir Charles Herbert Theophilus Metcalfe', *PICE*, 228 (1928), pp. 350–2, on p. 352.

101. Marsden and Smith, *Engineering Empires*, pp. 254–8. See also discussion in the Conclusion of this volume.

102. Ball, *The Law Affecting Engineers*, p. 4.

103. Alexander B. W. Kennedy, quoted in ibid., p. 3.

104. The policy was officially adopted at a council meeting in 1894 but it had been enforced prior to this date. See 'Special Council Meeting', 7 December 1894, CMB, 14, p. 714.

105. For examples, see AMICE M. H. Andersen to Council, 1 February 1898, CMB, 15, p. 20; AMICE John Hayes, AMICE R. E. Phillips, AMICE J. P. Baylis, AMICE W. H Severin to Council, 8 November 1904, CMB, 17, pp. 288–9.

106. Towards the end of the period the council had greater difficulties controlling this and therefore had to resort to formal means of control. In 1910 the previously unwritten rules were included into the by-laws and a permanent committee of professional conduct was set up which among other issues dealt with incidences of members advertising professional services. This was prompted by a rise in the number of engineers who violated this code. See in particular 'Report of Professional Conduct Committee', 23 February 1910, CMB, 19, pp. 393–4. In spite of this new departure there was still dissatisfaction among members that the ICE did not efficiently prevent touting. When the Association of Consulting Engineers was founded in 1912 by dissenters from the ICE it was openly stated that a main reason for establishing a new association for consulting engineers was that the ICE had not prevented the touting of consultancy services, in particular in the burgeoning field of electrical engineering. See Woodrow, *Tales of Victoria Street*, pp. 2–12.

107. Marsden and Smith, *Engineering Empires*, p. 239.

108. Arapostathis, 'Morality, Locality and "Standardization"'.

6 Vandals and Civilizers in Aswan and London

1. H. H. Statham, 'Philae and the Reservoir', *Builder* (3 March 1894), pp. 165–6, on p. 166. 'Timeo Danaos et Dona ferentes' translates as 'I fear the Greeks even if they bring gifts'.

2. The conflicting understandings are labelled Muscular Modernization and Paternalistic Preservation in C. Andersen, 'The Philae Controversy – Muscular Modernization and Paternalistic Preservation in Aswan and London', *History and Anthropology*, 22:2

(2011), pp. 203–20, which places the Philae controversy in the context of global heritage politics.

3. D. Cannadine, *Ornamentalism: How the British Saw their Empire* (Oxford: Oxford University Press, 2002), p. 19.

4. There are several good histories of the Nile and its changing uses. See, among others, R. O. Collins, *The Waters of the Nile* (Oxford: Clarendon Press, 1990); and T. Tvedt, *The River Nile in the Age of the British: Political Ecology and the Quest for Economic Power* (London: I. B. Tauris & Co., 2004).

5. For an introduction and overview, see Sayyid-Marsot, 'The British Occupation of Egypt'. Sir Evelyn Baring became Lord Cromer in 1892. For a fine biography, see Owen, *Lord Cromer*.

6. P. J. Cain 'Character and Imperialism: The British Financial Administration of Egypt, 1878–1914', *Journal of Imperial and Commonwealth History*, 34:2 (2006), pp. 177–200, on pp. 177–80. Many of Cain's points are anticipated by Roger Owen in a superb piece of intellectual history on 'the contemporary analysis of Egypt's financial situation'. See R. Owen, *Cotton and the Egyptian Economy, 1820–1914: A Study in Trade and Development* (Oxford: Oxford University Press, 1969), pp. 333–52.

7. The most comprehensive analysis remains R. L. Tignor, *Modernisation and British Colonial Rule in Egypt 1882–1914* (Princeton, NJ: Princeton University Press 1966).

8. Owen, *Cotton and the Egyptian Economy*, pp. 333–5; Tignor, *Modernisation and British Colonial Rule*, pp. 110–20.

9. See Tignor, 'The "Indianisation" of the Egypt Administration', pp. 654–7, for the Anglo-Indian engineers.

10. M. Hollings, *The Life of Sir Colin Scott-Moncrieff* (London: John Murray, 1917); C. Scott-Moncrieff, 'Irrigation in Egypt', *Nineteenth Century*, 17:2 (1885), pp. 343–53.

11. W. Willcocks, *Sixty Years in the East* (London: William Blackwood & Sons, 1935). For a study of Willcocks's turbulent career from a perspective of environmental history, see Beinart and Hughes, *Environment and Empire*, pp. 130–48.

12. J. C. Ross, 'Irrigation and Agriculture in Egypt', *Scottish Geographical Magazine*, 9 (1893), pp. 170–93.

13. E. Baring, Earl of Cromer, *Modern Egypt*, 2 vols (London: Macmillan, 1908), vol. 2, p. 465.

14. W. Willcocks, *Egypt Fifty Years Hence* (Cairo: Khedival Agricultural Society, 1902), p. 3.

15. W. Willcocks, *Egyptian Irrigation* (London: Millhouse, 1899), front page.

16. For two important recent studies of ideas about ancient Egypt in nineteenth-century British culture, see G. P. Nash, *From Empire to Orient: Travellers to the Middle East 1830–1926* (London: I. B. Tauris & Co., 2005); and D. Gange, 'Religion and Science in Late Nineteenth-Century British Egyptology', *Historical Journal*, 49:4 (2006), pp. 1083–102.

17. E. Said, *Orientalism* (New York: Pantheon Books, 1978). For an overview of the prolific literature on orientalism see, A. L. Macfie (ed.), *Orientalism – A Reader* (Edinburgh: Edinburgh University Press, 2000). For a well-founded reproach of uncritical use of Said's perspectives in the historiography of imperialism, see Washbrooke, 'Orients and Occidents'.

18. Buchanan, *The Engineers*, p. 150.

19. See, for example, G. F. Zimmer, *Engineering of Antiquity and Technical Progress in Arts and Crafts* (London: Probsthain & Co., 1913); J. Mansergh, 'Presidential Address', *PICE*, 143 (1900), pp. 2–83, on pp. 10–12.

20. C. Wood, *Olympian Dreamers: Victorian Classical Painters* (London: Constable, 1983), pp. 137–8.
21. There are a number of popular historical accounts of how the obelisk came to London, for example R. A. Hayward, *Cleopatra's Needles* (Thrapston: Moorland Publishing, 1978), which was published for the centenary of the event. Several of the individuals involved in the venture published their version of the story. The best for detail is J. E. Alexander, *Cleopatra's Needle* (London: Chatto & Windus, 1879).
22. Mackay, *The Life of Sir John Fowler*, pp. 266–8.
23. 'Obituary: John Dixon', *PICE*, 104 (1891), pp. 309–11.
24. For description of the design of the vessel, see 'The Cleopatra', *Engineer*, 54 (30 March 1877), pp. 342–5.
25. Baker, 'Cleopatra's Needle', pp. 234–6.
26. E. Wilson, *Our Egyptian Obelisk: Cleopatra's Needle* (London: Brain & Co., 1877), p. 2.
27. The events in the Bay of Biscay cleared the front page in several newspapers, for example *Illustrated London News* (27 October 1877).
28. For a lengthy coverage of the ceremony, see, 'Cleopatra's Needle', *The Times*, 13 September 1878.
29. 'Plaque for the Cleopatra's Needle', *Builder* (25 August 1878), p. 3.
30. Brown, *History of the Barrage*; Tignor, 'British Agricultural and Hydraulic Policy'.
31. Owen, *Cotton and the Egyptian Economy*, pp. 183–203.
32. W. Willcocks (with a note by W. E. Garstin), 'Report on Perennial Irrigation and Flood Protection for Egypt 1894', Archive of the ICE, no. 626.86 (62). For a good historical account of the construction of the first Aswan Dam, see N. A. F. Smith, *The Centenary of the Aswan Dam 1902–2002* (London: Thomas Telford, 2002).
33. Willcocks, 'Report on Perennial Irrigation', pp. 11–17. For the detailed maps accompanying the report, see W. Willcocks, 'Perennial Irrigation and Flood Protection for Egypt. Maps' (1894), Foreign Office Papers, FO 925/3008.
34. Willcocks, 'Report on Perennial Irrigation', pp. 4–5.
35. The political instability of Egypt also influenced this decision and the Philae controversy generally. In February 1894 an escalating crisis between the British and Khedive Abbas climaxed in the 'frontier incident' which brought out in the open that Egypt *de facto* was a British protectorate. This meant that the British administration had to adopt a defensive line during the Philae controversy in order not to jeopardize their position in domestic Egyptian politics as well as in international diplomacy. For an analysis of the political priorities of the British administration in 1894, see Owen, *Lord Cromer*, pp. 280–98.
36. W. Garstin, 'Note upon the Reports of the Technical Commission', in 'Reports of the Technical Commission', pp. vi–vii.
37. M. Boulé, 'Reservoirs in the Nile Valley. Report of M. Boulé (April 18, 1894)', in 'Reports of the Technical Commission', pp. 48–9.
38. Ibid., p. 34.
39. G. Torricelli and B. Baker, 'Report of Sir Benjamin Baker and of Mr G Torricelli on Question 5 of the Note of Under Secretary of State (April 10, 1894)', in 'Reports of the Technical Commission', pp. 5, 23.
40. W. Garstin, 'Note upon the Reports of the Technical Commission', in 'Reports of the Technical Commission', p. xvii.
41. Willcocks, 'Report on Perennial Irrigation', p. 6.

42. Cromer to Kimberley, 'Confidential Report' 15 July 1894, 'Irrigation of the Nile Valley Scheme of Cope Whitehouse. Raijan Reservoir etc.', vol. 1: 1888–94, Foreign Office Papers, FO 78/5261, no. 154, p. 3.
43. A. Edwards, *A Thousand Miles up the Nile* (London: Longmans, 1877), p. 227.
44. T. G. H. James (ed.), *Excavation in Egypt: The Egypt Exploration Society, 1882–1892* (London: British Museum Press, 1982). For a biography of the founder of the society, see B. Moon, *More Usefully Employed: Amelia B. Edwards, Writer, Traveller and Campaigner for Ancient Egypt* (London: Egypt Exploration Society, 2006). In 1919 the fund changed its name to the Egypt Exploration Society. It is still in existence today and remains the leading private organization for Egyptology in Britain. I would like to thank the Egypt Exploration Society for allowing me to make use of their archive and especially Patricia Spencer for her assistance during my time there in 2006–7.
45. W. H. Stiebing, *Uncovering the Past: A History of Archaeology* (Oxford: Oxford University Press, 1993), pp. 55–78; Gange, 'Religion and Science'.
46. M. S. Drower, 'The Early Years', in James (ed.), *Excavation in Egypt*, pp. 18–32, on pp. 29–32, provides a short account of the history of the SPMAE. For Poynter's work and career, see Wood, *Olympian Dreamers*, pp. 131–53. The archive of the SPMAE is kept at EES, box VIII: 'Minutes, Correspondence and Papers of the Society for the Preservation of Monuments of Ancient Egypt 1888–98'.
47. 'Committee for the Preservation of the Monuments of Ancient Egypt', undated revised circular, EES, box VIII, A, 1. In August 1888 the name of the organization was changed from 'Committee' to 'Society'.
48. 'Society for the Preservation of the Monuments of Ancient Egypt', printed pamphlet from the proceedings of the first meeting, EES, box VIII, B, 9, p. 1.
49. Ibid., p. 3.
50. 'Society for the Preservation of the Monuments of Ancient Egypt. First General Meeting', EES, box VIII, C, 27. Parts of the other side of this correspondence are in the Foreign Office Papers: 'Preservation of Ancient Monuments etc. 1894–1904', FO 78/5384. Substantial parts of the correspondence between Cromer and the Foreign Office on the Aswan Scheme and Philae have also survived in the Foreign Office Papers: 'Irrigation of the Nile Valley Scheme', FO 78/5261.
51. Julian Pauncefote [FO] to Poynter, 26 March 1889, EES, box VIII, A, 14. The society also directly contacted Cromer on the preservation issues, for example Henry Brackenbury [SPMAE] to Cromer, 16 March 1890, EES, box VIII, B, 17.
52. Cromer to Salisbury (Confidential Print), 7 March 1891, EES, box VIII, C, 18; Poynter to Salisbury, 11 May 1891, EES, box VIII, C, 24.
53. 'Second Annual Report of the Society for the Preservation of the Monuments of Ancient Egypt 1891', EES, box VIII, C, 27, pp. 3–4.
54. Moncrieff to SPMAE, 24 June 1890, EES, box VIII, B, 36.
55. G. H. Portal to Salisbury, 17 September 1890, EES, box VIII, B, 48; Ross to Poynter, 30 April 1892, EES, box VIII, D, 3.
56. Ross to Edwards, 27 September 1891, EES, box I, B, 1.
57. Edwards to Poynter, 15 October 1891, box I, B, 2.
58. Philip Currie [FO] to Poynter, 5 November 1891, box VIII, C, 31; Poynter to Edwards, 14 November 1891, box I, B, 3.
59. Edwards to Poynter, 26 November 1891, box VIII, C, 32.
60. Poynter to Edwards, 4 December 1891, box I, B, 5.
61. Morgay to Poynter, 16 February 1894, box VIII, D, 10.

62. E. Poynter, 'Letter to the Editor: Society for the Preservation of the Monuments of Ancient Egypt', *The Times*, 19 February 1894. 'Rien n'est sacré pour le sapeur' translates as 'Nothing is sacred to the sapper'.

63. Blunt to Poynter, 4 May 1894, EES, box VIII, D, 18.

64. *Reservoirs in the Valley of the Nile (with a map). Pamphlet Prepared for the Committee of the Society for the Preservation of the Monuments of Ancient Egypt*, EES, box VIII, D, 28. Blunt also directed attention to the fact that the Egyptian government would increase its land tax earnings substantially if large tracts of land in the Nile Valley were brought under perennial irrigation. This information was, however, not included in the pamphlet.

65. Ibid., pp. 4–6.

66. 'Lord Carlisle to Foreign Office: Memorial to the Honourable Earl of Kimberley Foreign Secretary of State', 15 June 1894, 'Preservation of Ancient Monuments', FO 78/5384, no. 124; 'Memorial Protesting against the Submersion of Philae', EES, box VIII, D, 20; Kimberley to Poynter 17 June 1894, EES, box VIII, D, 21.

67. 'Partition of the Egypt Exploration Fund', 'Preservation of Ancient Monuments', FO 78/5384, no. 234; 'Memorial for the Preservation of Philae' EES, box I, B, 18.

68. 'The Utilisation of the Nile', *Engineer*, 75 (2 March 1894), p. 172.

69. 'The Assuan Dam on the Nile', *Engineering* (20 April 1894), p. 521.

70. 'The Proposed Destruction of Philae', *Graphic* (10 March 1894), p. 279.

71. 'The Isles and Temple of Philae on the Upper Nile', *Illustrated London News* (17 March 1894), p. 331.

72. 'Fine Art Gossip', *Athenaeum* (3 March 1894), p. 282.

73. Statham, 'Philae and the Reservoir', p. 165.

74. J. C. Ross, 'Letter to the Editor: The Threatening Destruction of the Temple of Philae', *Academy* (3 February 1894), p. 110.

75. B. Baker, 'The Nile Reservoir', *Nineteenth Century*, 35:207 (1894), pp. 863–72, on pp. 863–4, 865–6, 871.

76. F. Dillon, 'The Proposed Nile Reservoir II: The Submergence of Philae', *Nineteenth Century*, 35:208 (1894), pp. 1019–25, on p. 1025.

77. Statham, 'Philae and the Reservoir', p. 166.

78. C. Scott-Moncrieff, 'The Nile', *Nature* (7 March 1895), pp. 444–6, on p. 446. The article was a printed version of lecture Scott-Moncrieff delivered at the Royal Institution on 25 January 1895.

79. F. L. Sandwith to Poynter, 26 June 1894, EES, box VIII, D, 22; W. Garstin, 'Note upon the Proposed Modification of the Assouan Dam', 'Irrigation of the Nile Valley Scheme', FO 78/5261, no. 165.

80. 'Society for the Preservation of the Monuments of Ancient Egypt', *Engineer*, 76 (15 February 1895), p. 154.

81. J. N. Lockyer, 'Perennial Irrigation in Egypt', *Nature* (24 May 1894), p. 80.

82. Garstin, 'Note upon the Proposed Modification of the Assouan Dam'.

83. W. Willcocks, *The Nile Reservoir Dam at Assuan, and After* (E. & F. N. Spon, 1903), p. 7.

84. Garstin, 'Note upon the Proposed Modification of the Assouan Dam'.

85. 'Egypt Exploration Fund, President Sir John Fowler', *The Times*, 14 August 1894.

86. Willcocks, *The Nile Reservoir Dam at Assuan*, pp. 7–9; Garstin, 'Note upon the Proposed Modification of the Assouan Dam', p. 5.

87. For unique photos of the partially submerged island of Philae during the first year after the dam had been completed, see photos taken by Baker and published in *Engineering* (23 January 1903), p. 114. In hindsight it is evident that the drowning of Philae was only

temporarily postponed. Within a decade arguments again gained momentum that the low-ered dam was inadequate for providing for the ever-growing freshwater needs in Egypt. In 1907–12 the dam was raised to the height originally intended which left the Philae temples underwater several months a year. This caused much damage to the structure and colour of the temples, but extensive and costly engineering works were, however, carried out to consolidate the foundation of the main buildings. For more on these works, see H. G. Lyons, *Ministry of Public Works: A Report on the Temples of Philae* (Cairo: National Printing Department, 1908). By then Lord Cromer had been elected president of the EEF and the Preservation Society had disbanded after Poynter was appointed curator of the Royal Academy. The dam was raised again in 1929–34. The temples of Philae were finally removed to the Island of Agilika in a campaign and programme led by UNESCO when the new Aswan High Dam was constructed in the 1950s to provide hydroelectricity for the growing population of Egypt. For an outline and firsthand account of the UNESCO pro-ject, see T. Säve-Söderbergh, *Temples and Tombs of Ancient Nubia: The International Rescue Campaign at Abu Simbel, Philae and Other Sites* (London: Thames & Hudson, 1987).

88. Buchanan, 'Gentleman Engineers', p. 424.
89. Fowler to Gosselin, 5 December 1887, EES, box XV, 5. The well-indexed internal cor-respondence among the EEF members is organized without the use of envelopes.
90. Fowler to Gosselin, 16 November 16 1889, EES, box X, I, 32; Gosselin to Fowler, 2 December 1889, box XX, B, 456. See also note 85 above.
91. Letter from E. M. Thompson, quoted in Mackay, *The Life of Sir John Fowler*, pp. 352–3.
92. Mackay, *The Life of Sir John Fowler*, pp. 9–11.
93. J. Fowler, lecture in Merchant Venturers' School in Bristol in 1893, quoted in ibid., p. 263.
94. Mackay, *The Life of Sir John Fowler*, p. 215.
95. Patterson to Fowler, 24 February 1894, EES, box XX, 809.
96. Fowler to Patterson, 12 March 1894, EES, box XV, 137.
97. Patterson to Fowler, 12 April 1894, EES, box XX, 825.
98. Fowler to Patterson, April 14 1894, EES, box XV, 112.
99. 'Sir John Fowler on the Philae Temples', *The Times*, 25 May 1894.
100. Patterson to Winlow, 26 May 1894, EES, box XX, 184.
101. John Fowler to Kimberley, 15 July 1894, 'Preservation of Ancient Monuments', FO 78/5384, no. 234.
102. P. Bourdieu, 'The Forms of Capital', in J. Richardson (ed.) *Handbook of Theory and Research for the Sociology of Education* (New York: Greenwood, 1986), pp. 241–58.
103. Baker, 'The Nile Reservoir', p. 865.
104. J. P. Mahaffy, 'The Destruction of Philae', *Nineteenth Century*, 35:208 (1894), pp. 1013–18, on p. 1016.

Conclusion

1. For critical discussions of 'the spokes in the wheel model', see, among others, Ballantyne, *Orientalism and Race*, pp. 14–15; and Magee and Thompson, *Empire and Globalisation*, pp. 17–19.
2. Mackenzie, '"The Second City of the Empire"'; Hazareesingh, 'Interconnected Synchro-nicities'; Forbes-Monroe, *Maritime Enterprise and Empire*.
3. Darwin, 'Imperialism and the Victorians', p. 640.
4. For an analysis of the Fowler–Whitton connection, see also Lee, *Colonial Engineer: John Whitton*; and Andersen, 'Colonial Connections'.

5.   Potter, 'Webs, Networks, and Systems', pp. 621–2, 645–6.
6.   Going, 'The Engineer and the Policy of Imperialism', p. 529.
7.   Buchanan, 'Gentlemen Engineers', pp. 407–8.
8.   Marsden and Smith, *Engineering Empires*, pp. 251–4.
9.   Thompson, *The Empire Strikes Back?*, pp. 11, 22–4.
10.   Cannadine, *Ornamentalism*, pp. 123–6.
11.   Rolt, *Victorian Engineering*, p. 163; Weiner, *English Culture*, pp. 41–67.
12.   Macleod, *Heroes of Invention*, p. 327.

WORKS CITED

Archival Material

Archive of the Institution of Civil Engineers

Advisory Committee for South Africa 1904–50, no. 189/107.

Advisory Committee: Transvaal Correspondence 1893–7, no. 189, box 20.

Coode Archive: Reports and Correspondence Selected from the Records of Coode & Partners, no. 627.2M.

Council Minute Books, nos 11–20, 1888–1913

Fowler, J., 'Egyptian Reports as the General Engineering Advisor to the Egyptian Government 1869–77', no. 62, M 603A.

Overseas Advisory Committees – General History from 1880, no. 187/185.

Technical Commission on Reservoirs, 'Reports of the Technical Commission on Reservoirs with a note by W. E. Garstin 1894', no. 627.8 (62) B.

Willcocks, W. (with a note by W. E. Garstin), 'Report on Perennial Irrigation and Flood Protection for Egypt 1894', no. 626.86 (62).

Colonial Office Papers, 372/245 [372: Confidential print, Africa South].

Egypt Exploration Society Archive

Archive of the Egypt Exploration Fund.

Archive of the Society for the Preservation of Monuments of Ancient Egypt.

Foreign Office Papers

'Irrigation of the Nile Valley Scheme of Cope Whitehouse. Raijan Reservoir etc.', vol. 1: 1888–94, FO 78/5261.

'Preservation of Ancient Monuments etc. 1894–1904', FO 78/5384.

Willcocks, W., 'Perennial Irrigation and Flood Protection for Egypt. Maps' (1894), FO 925/3008.

Parliamentary Papers

Command Papers; Accounts and Papers LXIII.611/C9331, 'Report on the Uganda Railway by Sir Guilford Lindsey Molesworth', at http://gateway.proquest.com/openurl?url_

ver=Z39.88-2004&res_dat=xri:hcpp&rft_dat=xri:hcpp:rec:1899-077292 [accessed 26 April 2011].

House of Commons Papers; Accounts and Papers LVI. 637/103, 'The Remuneration of the Consulting Engineers to Crown Colonies and Protectorates', at http://gateway.proquest.com/openurl?url_ver=Z39.88-2004&res_dat=xri:hcpp&rft_dat=xri:hcpp:rec:1907-007668 [accessed 26 April 2011].

Rhodes House Archive, Oxford

Girouard, P., 'Autobiography', MSS Afr. S. 1865.

Lugard, Lord, Papers of, MSS Afr. R. 231.

Patterson, J. H., Papers and Diaries of, MSS Afr. R. 93.

Seymour, A. R., 'Tropical Railway', MSS Afr. R. 213.

Shelford, F., Papers of, MSS Afr. S. 2193.

Whitehouse, B., 'Account of Survey of the Northern Section of Lake Victoria', MSS Afr. S. 1294.

Whitehouse, G., Papers of, MSS Afr. S. 1046.

Newspapers, Journals, Magazines

African and Eastern Engineering.

African Engineering.

African World – A Weekly Periodical of Africa.

Athenaeum.

Builder.

Engineer.

Engineering.

Engineering Magazine – An Industrial Review.

Evening Post.

Feilden's Magazine – The World's Record of Industrial Progress.

Financier.

Graphic.

Illustrated London News.

Journal of the Institution of Civil Engineers.

Journal of the Royal African Society.

Proceedings of the Institution of Civil Engineers (PICE).

The Times.

Published Sources

A & C's Who Was Who, 3 vols (London: Adam & Charles Black, 1916–40).

Adas, M., *Machines as the Measure of Men: Science, Technology, and Ideologies of Western Dominance* (Ithaca, NY: Cornell University Press, 1989).

Alexander, J. E., *Cleopatra's Needle* (London: Chatto & Windus, 1879).

Allen, C. E., 'Lest we Forget', *Feilden's Magazine*, 1:1 (1899), p. 4.

Andersen, C., 'The Philae Controversy – Muscular Modernization and Paternalistic Preservation in Aswan and London', *History and Anthropology*, 22:2 (2011), pp. 203–20.

—, 'Colonial Connections and Consulting Engineers 1850–1914', *Engineering History and Heritage*, 165:1 (forthcoming 2011).

—, 'Explorer-Engineers take the Field: Imperial Engineers, Africa and the Late-Victorian Public', in M. Harbsmeier, K. H. Nielsen and C. J. Ries (eds), *Scientists and Scholars in the Field: Studies in the History of Expeditions and Fieldwork* (Aarhus: Aarhus University Press, forthcoming 2011).

Andersen, C., and Hjermitslev, H. H., 'Directing Public Interest: Danish Newspaper Science 1900–1903', *Centaurus: An International Journal for the History of Science*, 51:2 (2009), pp. 143–67.

Appleyard, R. *The History of the Institution of Electrical Engineers 1871–1931* (London: Institution of Electrical Engineers, 1939).

Arapostathis, S., 'Consulting Engineers in the British Electric Light and Power Industry, c. 1880–1914' (unpublished PhD thesis, University of Oxford, 2006).

—, 'Morality, Locality and "Standardization" in the Work of the British Consulting Electrical Engineers, 1880–1914', in G. Gooday and J. Sumner (eds), *History of Technology*, 28, special issue: 'By Whose Standards? Standardization, Stability and Uniformity in the History of Information and Electrical Technologies' (2008), pp. 53–75.

Arnold, D., *Science, Technology and Medicine in Colonial India* (Cambridge: Cambridge University Press, 2000).

—, 'Europe, Technology and Colonialism in the 20th Century', *History and Technology*, 21:1 (2005), pp. 85–106.

Association of Consulting Engineers, *History of the Formation of the Association of Consulting Engineers* (London: Association of Consulting Engineers, 1930).

Atkins, P. J., *The Directories of London 1677–1977* (London: Continuum, 1990).

Bachmann-Medick, D., *Cultural Turns, Neuorientierungen in den Kulturwissenschaften* (Reinbek: Rowohlt, 2006).

Baker, B., 'Cleopatra's Needle', *PICE*, 62 (1878), pp. 233–44.

—, 'The Nile Reservoir', *Nineteenth Century*, 35:207 (1894), pp. 863–72.

—, 'Presidential Address', *PICE*, 123 (1896), pp. 1–38.

Baker, E. C. *Sir William Preece, Victorian Engineer Extraordinary* (London: Hutchinson & Co., 1976).

—, *Preece and Those who Followed: Consulting Engineers in the Twentieth Century* (Dublin: Reprographic, 1980).

Ball, M., and D. Sunderland, *An Economic History of London 1800–1914* (London: Routledge, 2000).

Ball, W. V., *The Law Affecting Engineers* (London: A. Constable, 1909).

Ballantyne, T., *Orientalism and Race: Aryanism in the British Empire* (Hampshire: Palgrave Macmillan, 2001).

Baring, E., Earl of Cromer, *Modern Egypt*, 2 vols (London: Macmillan, 1908).

Barrett, J., 'England, America, and Germany as Allies for the Open Door, I', *Engineering Magazine*, 17:6 (1899), pp. 893–902.

Barton, R., '"Huxley, Lubbock, and Half a Dozen Others": Professionals and Gentlemen in the Formation of the X Club 1851–1864', *Isis*, 89:3 (1998), pp. 410–44.

Basalla, G., 'The Spread of Western Science', *Science*, 156:3775 (1967), pp. 611–22.

Beckett, B. H., *Scientific London* (London: John Murray, 1875).

Beinart, W., and L. Hughes, *Environment and Empire* (Oxford: Oxford University Press, 2007).

Bhaba, H. K., *The Location of Culture* (London: Routledge, 1994).

Biagioli, M., *Galileo Courtier: The Practice of Science in the Culture of Absolutism* (Chicago, IL: Chicago University Press, 1993).

Birkett, D. J., *Mary Kingsley: Imperial Adventuress* (Hampshire: Palgrave Macmillan, 1992).

Bourdieu, P., 'The Forms of Capital', in J. Richardson (ed.), *Handbook of Theory and Research for the Sociology of Education* (New York: Greenwood, 1986), pp. 241–58.

Brian, D., *The Seven Lives of Colonel Patterson: How an Irish Lion Hunter Led the Jewish Legion to Victory* (New York: Syracuse, 2008).

Bowler, P. J., 'Experts and Publishers: Writing Popular Science in Early Twentieth-Century Britain', *British Journal for the History of Science*, 39:2 (2006), pp. 159–87.

Bradley S., and N. Pevsner, *London 6: Westminster* (London: Yale University Press, 2003).

Brooke, D., *The Railway Navvy: 'That Despicable Race of Men'* (Devon: Davis & Charles, 1983).

Brown, R. H., *History of the Barrage at the Head of the Delta of Egypt*, introductory note W. E. Garstin (Cairo: Public Works Print, 1896).

Buchanan, R. A., 'Gentlemen Engineers: The Making of a Profession', *Victorian Studies*, 26:4 (1983), pp. 407–31.

—, '"Institutional Proliferation" in the British Engineering Profession 1847–1914', *Economic History Review*, n.s., 38:1 (1985), pp. 42–60.

—, 'The Diaspora of British Engineering', *Technology and Culture*, 27:3 (1986), pp. 501–24.

—, *The Engineers: A History of the Engineering Profession in Britain 1750–1914* (London: Jessica Kingsley Publishers, 1989).

—, *Brunel: The Life and Times of Isambard Kingdom Brunel* (London: Continuum, 2002).

Burge, C. O., 'The Hawkesbury Bridge, New South Wales', *PICE*, 101 (1890), pp. 2–12.

—, *The Adventures of a Civil Engineer: Fifty Years on Five Continents* (London: Alston Rivers, 1909).

Burke, P., *A Social History of Knowledge: From Gutenberg to Diderot* (Cambridge: Polity, 2000).

Burton, A., *Burdens of History: British Feminists, Indian Women, and Imperial Culture, 1865–1915* (Chapel Hill, NC: University of North Carolina Press, 1994).

Cain, P. J., 'Character and Imperialism: The British Financial Administration of Egypt, 1878–1914', *Journal of Imperial and Commonwealth History*, 34:2 (2006), pp. 177–200.

Cain, P. J., and A. G. Hopkins, *British Imperialism 1688–2000* (London: Longman, 1993).

Cannadine, D., *Ornamentalism: How the British Saw their Empire* (Oxford: Oxford University Press, 2002).

—, 'Engineering History or the History of Engineering', *Transactions of the Newcomen Society*, 74B (2004), pp. 163–80.

Cantor, G., 'The Scientist as Hero: Public Images of Michael Faraday', in M. Shortland and R. Yeo (eds), *Telling Lives in Science: Essays on Scientific Biography* (Cambridge: Cambridge University Press, 1996), pp. 171–93.

Cantor, G., G. Dawson, G. Gooday, R. Noakes, S. Shuttleworth and J. Topham, *Reading the Magazine of Nature: Science in the Nineteenth-Century Periodical* (Cambridge: Cambridge University Press, 2004).

Carlyle, T., *Sartor Resartus* (1831; Oxford: Oxford University Press, 1999).

Chatzis, K., 'Introduction: The National Identities of Engineers', *History and Technology*, 23:3 (2007), pp. 193–6.

Chew, K., and A. Wilson, *Victorian Science and Engineering Portrayed in the Illustrated London News* (London: Allan Sutton, 1993).

Chrimes, M., 'John Fowler: Engineer or Manager?', *PICE*, 97 (1993), pp. 135–43.

—, 'British Civil Engineers in Canada 1830–90', *Proceedings of CSCE International Symposium on Civil Engineering History*, 1 (2005), pp. 3–28.

—, 'British and Irish Civil Engineers in the Development of Argentina in the Nineteenth Century', in M. Dunkeld et al. (eds), *Proceedings of the Second International Congress on Construction History*, 1 (2006), pp. 675–94.

Churchill, W., *My African Journey* (London: Hodder & Stoughton, 1908).

Collins, R. O., *The Waters of the Nile* (Oxford: Clarendon Press, 1990).

Coode, J., 'Presidential Address', *PICE*, 99 (1889), pp. 1–40.

Cooper, F., *Colonialism in Question: Theory, Knowledge, History* (Berkeley, CA: University of California Press, 2005).

Cowell, F. R., *The Athenaeum: Club and Social Life in London 1824–1974* (Oxford: Heinemann, 1975).

Croxton, A. H., *Railways of Rhodesia: The Story of the Beira, Mashonaland and Rhodesian Railways* (Devon: David & Charles, 1974).

Crozier, A., 'Sensationalising Africa: British Medical Impressions of Sub-Saharan Africa 1890–1939', *Journal of Imperial and Commonwealth History*, 35:3 (2007), pp. 393–415.

Darnton, R., *The Kiss of Lamourette: Reflections in Cultural History* (Markham: Penguin, 1990).

Darwin, J., 'Imperialism and the Victorians: The Dynamics of Territorial Expansion', *English Historical Review*, 112:447 (1997), pp. 614–42.

—, *After Tamerlane: The Global History of Empire since 1405* (London: Penguin, 2007).

—, *The Empire Project: The Rise and Fall of the British World-System, 1830–1970* (Cambridge: Cambridge University Press, 2009).

Davenport-Hines, R. P. T., 'The British Engineers' Association and Markets in China 1900–1930', in R. P. T. Davenport-Hines (ed.), *Markets and Bagmen: Studies in the History of Marketing and British Industrial Performance, 1830–1939* (London: Gower Publishing, 1987), pp. 102–30.

Dickens, C., *Little Dorrit* (1857; New York: Digireads, 2009).

Dictionary of South African Biography, 5 vols (Pretoria: Nasionale Boekhandel; Durban: Butterworth & Co., 1968–87).

Dillon, F. 'The Proposed Nile Reservoir II: The Submergence of Philae', *Nineteenth Century*, 35:208 (1894), pp. 1019–25.

Driver, F., *Geography Militant: Cultures of Exploration and Empire* (Oxford: Blackwell Publishers, 2001).

Driver F., and D. Gilbert, 'Imperial Cities: Overlapping Territories, Intertwined Histories', in Driver and Gilbert (eds), *Imperial Cities*, pp. 2–17.

— (eds), *Imperial Cities: Landscape, Display and Identities* (Manchester: Manchester University Press, 1999).

Driver, F., and L. Martins (eds), *Tropical Visions in an Age of Empire* (Chicago, IL: Chicago University Press, 2005).

Drayton, R., *Nature Government: Kew Gardens and the Improvement of the World* (London: Yale University Press, 2000).

Drower, M. S., 'The Early Years', in James (ed.), *Excavation in Egypt*, pp. 18–32.

Dubow, S., 'A Commonwealth of Science: The British Association in South Africa, 1905 and 1929', in S. Dubow (ed.), *Science and Society in Southern Africa* (Manchester: Manchester University Press, 2000), pp. 66–100.

—, *A Commonwealth of Knowledge: Science, Sensibility, and White South Africa 1820–2000* (Oxford: Oxford University Press, 2006).

Dunkeld, M., 'Paraphernalia of the Professions – The Case of the Institution of Civil Engineers' (unpublished manuscript).

Dunlap, J. R., 'Editorial Comment: Our Tenth Anniversary', *Engineering Magazine*, 21:1 (1901), p. 114.

Edgerton, D., *Science, Technology and the British 'Decline' 1870–1970* (Cambridge: Cambridge University Press, 1996).

—, *The Shock of the Old: Technology and Global History since 1900* (New York: Oxford University Press, 2006)

Edwards, A., *A Thousand Miles up the Nile* (London: Longmans, 1877).

Fage, J. D., 'When the African Society was Founded: Who were the Africanists?', *African Affairs*, 94:376 (1995), pp. 369–81.

Feilden, T., 'No Apology', *Feilden's Magazine – The World's Record of Industrial Progress*, 1:1 (1899), p. 1.

Ferguson, E. S., 'Technical Journals and the History of Technology', in S. H. Cutcliffe and R. C. Post (eds), *In Context: History and the History of Technology, Essays in Honour of Melvin Kranzberg* (New York: Lehigh University Press, 1989), pp. 53–71.

Fidler, T. C., *Civil Engineering* (London: Methuen & Co., 1905).

Flint, J. E., 'Britain and the Scramble for Africa', in Winks (ed.), *The Oxford History of the British Empire, Vol. V*, pp. 450–63.

Forbes-Monroe, J., *Maritime Enterprise and Empire: Sir William Mackinnon and His Business Network, 1823–1893* (London: Boydell Press, 2003).

Fowler J., and B. Baker, 'A Sweet-Water Ship Canal through Egypt', *Nineteenth Century*, 13:1 (1883), pp. 166–72.

Fox, F., *River, Road, and Rail: Some Engineering Reminiscences* (London: John Murray, 1904).

—, *Sixty-Three Years of Engineering, Scientific and Social Work* (London: John Murray, 1924).

Fukuyama, F., *Trust: The Social Virtues and the Creation of Prosperity* (London: Penguin, 1996).

Fyfe, A., and B. Lightman (eds), *Science in the Marketplace: Nineteenth-Century Sights and Experiences* (Chicago, IL: Chicago University Press, 2007).

Galbraith, J. S., *Mackinnon and East Africa, 1875–1895: A Study in the 'New Imperialism'* (Cambridge: Cambridge University Press, 1972).

—, *Crown and Charter – The Early Years of British South Africa Company* (Berkeley, CA: University of California Press, 1974).

Gallagher, J., and R. E. Robinson, 'The Imperialism of Free Trade', *Economic History Review*, 2nd series, 6:1 (1953), pp. 1–15.

Gange, D., 'Religion and Science in Late Nineteenth-Century British Egyptology', *Historical Journal*, 49:4 (2006), pp. 1083–102.

Girouard, P., 'The Railways of Africa', *Scribner's Magazine – Monthly with Illustrations*, 39 (1906), pp. 553–69.

Going, C. B., 'The Engineer and Imperialism', *Engineering Magazine*, 16:4 (1899), pp. 19–25.

—, 'The Engineer and the Policy of Imperialism', *Engineering Magazine*, 16:4 (1899), pp. 527–32.

—, 'An Editorial Review: The Industrial Significance of the Anglo-German Alliance', *Engineering Magazine*, 20:3 (1901), pp. 325–31.

Gooday, G., 'Lies, *Damned Lies and Declinism*: *Lyon Playfair, the Paris 1867 Exhibition and the Contested Rhetorics of Scientific Education and Industrial Performance*, in I. Inkster (ed.), *The Golden Age*: *Essays in British Social and Economic History, 1850–70* (*Aldershot*: Ashgate, 2000), pp. 105–21.

—, *Domesticating Electricity: Technology, Uncertainty and Gender, 1880–1914* (London: Pickering & Chatto, 2008).

—, 'Liars, Experts and Authorities', *History of Science*, 46:4 (2008), pp. 431–56.

Gunston, H., 'The Planning and Construction of the Uganda Railway', *Journal of Newcomen Society*, 74A (2004), pp. 45–71.

Hardy, R., *The Iron Snake* (Glasgow: Collins, 1965).

Hargreaves, J. D., *West Africa Partitioned*, 2 vols (London: Macmillan, 1974–85).

Hawkshaw, J., *Some Reminiscences of South America: From Two and a Half Years' Residence in Venezuela* (London: Jackson & Walford, 1838).

Hayward, R. A., *Cleopatra's Needles* (Thrapston: Moorland Publishing, 1978).

Hazareesingh, S., 'Interconnected Synchronicities: The Production of Bombay and Glasgow as Modern Global Ports c.1850–1880', *Journal of Global History*, 4:1 (2009), pp. 7–31.

Headrick, D. R., *The Tools of Empire: Technology and European Imperialism in the Nineteenth Century* (New York: Oxford University Press, 1981).

—, *The Tentacles of Progress: Technology Transfer in the Age of Imperialism, 1850–1940* (Oxford: Oxford University Press, 1988).

—, *Power over Peoples: Technology, Environments, and Western Imperialism, 1400 to the Present* (Princeton, NJ: Princeton University Press, 2010).

Hill, M. F., *The Permanent Way: The Story of the Kenya and Uganda Railway* (Nairobi: East African Railways, 1957).

Hobson, G. A. 'The Great Zambezi Bridge', in Weinthal (ed.), *The Story of the Cape to Cairo Railway*, vol. 1, pp. 43–59.

Hoffenberg, P. H., *An Empire on Display: English Indian and Australian Exhibitions from Crystal Palace to the Great War* (Berkeley, CA: University of California Press, 2001).

Hollings, M., *The Life of Sir Colin Scott-Moncrieff* (London: John Murray, 1917).

Hopkins, A. G., 'Explorers' Tales: Stanley Presumes – Again', *Journal of Imperial and Commonwealth History*, 36:4 (2008), pp. 669–84.

Hunter, F. R., *Egypt under the Khedives, 1805–1879: From Household Government to Modern Bureaucracy* (Cairo: American University Press in Cairo, 1999).

Høeg, P., 'Journey into a Dark Heart', in P. Høeg, *Tales of the Night* (1990; London: Penguin, 1999), pp. 3–38.

Institution of Civil Engineers, *Charter, Supplemental Charter, By-Laws and Lists of Members 1885* (London: Institution of Civil Engineers, 1885).

—, *Charter, Supplemental Charter, By-Laws and Lists of Members 1887* (London: Institution of Civil Engineers, 1887).

—, *Charter, Supplemental Charter, By-Laws and Lists of Members 1895* (London: Institution of Civil Engineers, 1895).

—, *Charter, Supplemental Charter, By-Laws and Lists of Members 1896* (London: Institution of Civil Engineers, 1896).

—, *Charter, Supplemental Charter, By-Laws and Lists of Members 1897* (London: Institution of Civil Engineers, 1897).

—, *Charter, Supplemental Charter, By-Laws and Lists of Members 1898* (London: Institution of Civil Engineers, 1898).

—, *Charter, Supplemental Charter, By-Laws and Lists of Members 1902* (London: Institution of Civil Engineers, 1902).

—, *Report of the Proceedings at the Ceremony of Laying the Foundation Stone of the New Building on Tuesday, the 25th October, 1910* (London: Institution of Civil Engineers, 1911).

—, *Charter, Supplemental Charter, By-Laws and Lists of Members 1912* (London: Institution of Civil Engineers, 1912).

—, *A Brief History of the Institution of Civil Engineers with an Account of the Charter Centenary Celebration June 1928* (London: Institution of Civil Engineers, 1928).

Jackson, F., *Early Days in East Africa* (London: Collins, 1931).

Jaekel, F., *History of the Nigerian Railway*, 3 vols (Ibadan: Spectrum Books, 1997).

James, T. G. H. (ed.), *Excavation in Egypt: The Egypt Exploration Society, 1882–1892* (London: British Museum Press, 1982).

Jarvis, A., *Samuel Smiles and the Construction of Victorian Values* (Gloucestershire: Sutton Publishing, 1997).

Joby, R. S., *The Railway Builders: Lives and Works of the Victorian Railway Contractors* (London: David & Charles, 1983).

Johnson, J. A., 'Germany: Discipline – Industry – Profession: German Chemical Organisations', in A. K. Nielsen and S. Štrbáňová, *Creating Networks in Chemistry: The Founding and Early History of Chemical Societies in Europe* (London: RSC Publishing, 2008), pp. 113–38.

Katzenellenbogen, S., *Railways and the Copper Mines of Katanga* (Oxford: Oxford University Press, 1973).

Keene, D., 'Cities and Empires', *Journal of Urban History*, 32:1 (2005), pp. 8–21.

Kelly's Directories of the Engineers, Iron and Metal Trades 1883 (London: Kelly & Co., 1883).

Kelly's Directories of the Engineers, Iron and Metal Trades 1890 (London: Kelly & Co., 1890).

Kelly's Directories of the Engineers, Iron and Metal Trades 1901 (London: Kelly & Co., 1901).

Kelly's Directories of the Engineers, Iron and Metal Trades 1909 (London: Kelly & Co., 1909).

Kelly's Mayfair, St. James's, Soho and Westminster Directory 'Buff Book' 1887 (London: Kelly & Co., 1887).

Kennedy, D., 'Imperial History and Post-Colonial Theory', *Journal of Imperial and Common-wealth History*, 24:3 (1996), pp. 345–63.

—, 'British Exploration in the Nineteenth Century: A Historiographical Survey', *History Compass*, 5:6 (2007), pp. 1879–900.

Ketchum, H. G. C., *The Chignecto Ship Railway* (Boston, MA: Damrell & Upham, 1893).

Kirk-Greene, A., 'Canada in Africa: Sir Percy Girouard, Neglected Colonial Governor', *African Affairs*, 83:331 (1984), pp. 207–39.

Krugman, P., *Geography and Trade* (Boston, MA: MIT Press, 1991).

Kubicek, R., 'British Expansion, Empire and Technological Change', in Porter (ed.), *The Oxford History of the British Empire, Vol. III*, pp. 258–77.

Laidlaw, Z., *Colonial Connections, 1815–45: Patronage, the Information Revolution and Colonial Government* (Manchester: Manchester University Press, 2005).

Lambert, D., and A. Lester (eds), *Colonial Lives Across the British Empire: Imperial Careering in the Long Nineteenth Century* (Cambridge: Cambridge University Press, 2005).

—, 'Introduction: Imperial Spaces, Imperial Subjects', in Lambert and Lester (eds), *Colonial Lives Across the British Empire*, pp. 1–31.

Lane, M. R., *The Rendel Connection: A Dynasty of Engineers* (London: Quiller Press, 1989).

Ledger, S., and R. Luckhurst, *The Fin de Siècle: A Reader in Cultural History c. 1880–1900* (Oxford: Oxford University Press, 2000).

Lee, R. *The Greatest Public Work: The New South Wales Railways, 1848–1889* (Sydney: Southwood Press, 1988).

—, *Colonial Engineer: John Whitton 1819–1898 and the Building of Australia's Railways* (Sydney: University of New South Wales Press, 2000).

Lefebvre, H., 'The Other Parises', in *Henri Lefebvre: Key Writings*, ed. S. Elden, E. Lebas and E. Kofman (London: Continuum, 2003), pp. 151–8.

Lester, A., *Imperial Networks: Creating Identities in Nineteenth-Century South Africa and Britain* (London: Routledge, 2001).

—, 'Imperial Circuits and Networks: Geographies of the British Empire', *History Compass*, 4:1 (2006), pp. 124–41.

Lewis, C., *Clara in Blunderland*, pictures by S. R. (London: William Heinemann, 1902).

Livingstone, D. N., *Putting Science in its Place – Geographies of Scientific Knowledge* (Chicago, IL: Chicago University Press, 2003).

Lockyer, J. N., 'Perennial Irrigation in Egypt', *Nature* (24 May 1894), p. 80.

Lunn, J., 'The Political Economy of Primary Railway Construction in Rhodesia', *Journal of African History*, 33:2 (1992), pp. 239–54.

Lyons, H. G., *Ministry of Public Works: A Report on the Temples of Philae* (Cairo: National Printing Department, 1908).

Macfie, A. L. (ed.), *Orientalism – A Reader* (Edinburgh: Edinburgh University Press, 2000).

Mackay, T. *The Life of Sir John Fowler – Engineer* (London: John Murray, 1900).

Mackenzie, J. M. (ed.), *Imperialism and Popular Culture* (Manchester: Manchester University Press, 1986).

—, *Propaganda and Empire: Manipulation of British Public Opinion, 1880–1960* (Manchester: Manchester University Press, 1986).

—, '"The Second City of the Empire": Glasgow – Imperial Municipality', in Driver and Gilbert (eds), *Imperial Cities*, pp. 215–37.

—, 'The Iconography of the Exemplary Life: The Case of David Livingstone', in G. Cubitt and A. Warren (eds), *Heroic Reputations and Exemplary Lives* (Manchester: Manchester University Press, 2000), pp. 84–104.

—, 'Comfort and Conviction: A Response to Bernard Porter', *Journal of Imperial and Commonwealth History*, 36:4 (2008), pp. 659–68.

Macleod, C., 'The Nineteenth Century Engineer as Cultural Hero', in A. Kelly and M. Kelly (eds), *Brunel: In Love with the Impossible* (Reston, VA: American Society of Civil Engineers, 2006), pp. 61–79.

—, *Heroes of Invention: Technology, Liberalism and British Identity, 1750–1914* (Cambridge: Cambridge University Press, 2007).

Macleod, C., and A. Nuvolari, 'The Pitfalls of Prosopography: Inventors in the Dictionary of National Biography', *Technology and Culture*, 47:4 (2006), pp. 757–76.

Magee G. B., and A. Thompson, *Empire and Globalisation: Networks of People, Goods and Capital in the British World, c.1850–1914* (Cambridge: Cambridge University Press, 2010).

Mahaffy, J. P., 'The Destruction of Philae', *Nineteenth Century*, 35:208 (1894), pp. 1013–18.

Mansergh, J., 'Presidential Address', *PICE*, 143 (1900), pp. 2–83.

Marrison, A, 'Comments on Howe and Sykes', in A. Marrison (ed.), *Free Trade and its Reception, 1815–1960* (London: Routledge, 1998), pp. 203–7.

Marsden B., and C. Smith, *Engineering Empires: A Cultural History of Technology in Nineteenth-Century Britain* (Hampshire: Palgrave Macmillan, 2007).

Marshall, A., *The Economics of Industry* (London: John Murray, 1888).

Masseys, D., *Space, Place and Gender* (Minneapolis, MN: University of Minnesota Press, 1994).

Maylam, P., 'The Making of the Kimberley–Bulawayo Railway', *Rhodesian History*, 8 (1977), pp. 13–35.

—, *Rhodes, the Tswana, and the British: Colonialism, Collaboration and Conflict in the Bechuanaland Protectorate 1885–1899* (London: Greenwood Press, 1980).

Mazlich, B., 'Technology and Social Relations: From Patronage to Networks', in P. Lyth and H. Trischler (eds), *Wiring Prometheus: Globalisation, History and Technology* (Aarhus: Aarhus University Press, 2004), pp. 21–35.

Merrington, P., 'A Staggered Orientalism: The Cape to Cairo Imagery', *Poetics Today*, 22:2 (2001), pp. 323–64.

Metcalf, T. R., *Imperial Connections: India in the Indian Ocean Arena 1860–1920* (Los Angeles, CA: University of California Press, 2007).

Metcalfe, C., *The Cape-to-Cairo Line: How the War Might Solve a Problem* (London: Royal Empire Society, 1914).

—, 'Railway Development of Africa, Present and Future', *Geographical Journal*, 47:1 (1916), pp. 3–17.

—, 'My Story of the Scheme', in Weinthal (ed.), *The Story of the Cape to Cairo Railway*, vol. 1, pp. 91–9.

Metcalfe, C., and F. I. Richarde-Seaver, 'The British Sphere of Influence in South Africa', *Fortnightly Review*, 267 (1889), pp. 351–63.

Michie, R., *The City of London: Continuity and Change since 1850* (Hampshire: Palgrave Macmillan, 1992).

Miller, C., *The Lunatic Express – An Entertainment in Imperialism* (London: Penguin, 1977).

Misa, T. J., *Leonardo to the Internet: Technology and Culture from the Renaissance to the Present* (Baltimore, MD: Johns Hopkins University Press, 2004).

Mitchell, L., 'The Cape to Cairo Railway', *Journal of the Society of Arts*, 55:2822 (1906), pp. 98–109.

Molesworth, E. J., *Life of Sir Guildford Molesworth: The Nestor of the Engineering Profession* (London: E. & F. N. Spon, 1922).

Molesworth, G. L., *Imperialism and Free Trade* (London: E & F. N. Spon, 1886).

—, *Democracy and War* (London: E. & F. N. Spon, 1889).

—, *Our Empire under Protection and Free Trade* (London: Lock & Co., 1902).

—, *Economic and Fiscal Fallacies* (London: Longman, 1910).

Moon, B., *More Usefully Employed: Amelia B. Edwards, Writer, Traveller and Campaigner for Ancient Egypt* (London: Egypt Exploration Society, 2006).

Morris, S., 'Indians in East Africa: A Study in a Plural Society', *British Journal of Sociology*, 7:3 (1956), pp. 194–211.

Mortimer, J., *Zerah Colburn: Spirit of Darkness* (Bury St Edmunds: Arima, 2005).

Murray, T., 'Thomas Stewart – First South African Consulting Engineer', *Documents of the Association of Consulting Engineers*, at http://email.asce.org/international/documents/05.MurrayStewartTWoodhead.pdf [accessed 30 November 2010].

Nash, G. P., *From Empire to Orient: Travellers to the Middle East 1830–1926* (London: I. B. Tauris & Co., 2005).

Naval Intelligence Department, *Handbook of Railways in Africa* (London: Admiralty, 1919).

North, S. J., *Europeans in British Administrated East Africa: A Biographical Listing* (London: S. J. North, 2005).

Owen, N., 'Critics of Empire in Britain', in J. M. Brown and W. R. Louis (eds), *The Oxford History of the British Empire; Vol. IV: The Twentieth Century* (Oxford: Oxford University Press, 1999), pp. 188–212.

Owen, R., *Cotton and the Egyptian Economy, 1820–1914: A Study in Trade and Development* (Oxford: Oxford University Press, 1969).

—, *Lord Cromer – Victorian Imperialist, Edwardian Proconsul* (Oxford: Oxford University Press 2004).

Paish, G., 'Great Britain's Capital Investment in Other Lands', *Journal of the Royal Statistical Society*, 72:3 (1909), pp. 465–95.

Parsons, R. H., *A History of the Institution of Mechanical Engineers 1847–1947* (London: Institution of Mechanical Engineers, 1947).

Patterson, J. H., *The Man-Eaters of Tsavo and Other East African Adventures* (1907; New York: Hard Press, 2003).

Pendred, B. (ed.), *The Engineer. Centenary Number: A Study of Influences of Engineering Advancement 1856–1956* (London: Morgan Brothers, 1956).

Perkin, H., *The Rise of Professional Society: England since 1880* (London: Routledge, 1989).

Phimister, I. R., 'Rhodes, Rhodesia and the Rand', *Journal of Southern African Studies*, 1:1 (1974), pp. 74–91.

—, 'Towards a History of Zimbabwe's Railways', *Zimbabwean History*, 12 (1981), pp. 61–91.

Picon, A., 'Engineers and Engineering History: Problems and Perspectives', *History and Technology*, 20:4 (2004), pp. 421–36.

Pieterse, J. N., 'Globalization as Hybridization', in M. Featherstone, S. Lash and R. Robertson (eds), *Global Modernities* (London: Sage Publications, 1995), pp. 45–67.

Port, M. H., *Imperial London: Civil Government Building in London 1851–1914* (London: Yale University Press, 1995).

Porter, A. (ed.), *The Oxford History of the British Empire, Vol. III: The Nineteenth Century* (Oxford: Oxford University Press, 1999).

Porter, B., *The Absent-Minded Imperialists: Empire, Society and Culture in Britain* (Oxford: Oxford University Press, 2004).

—, *The Lion's Share: A Short History of British Imperialism*, 4th edn (London: Longman, 2004).

Porter, D. H., and G. C. Clifton, 'Patronage, Professional Values and Victorian Public Works: Engineering and Contracting the Thames Embankment', *Victorian Studies*, 31:3 (1988), pp. 321–49.

Porter, M. E., *The Competitive Advantages of Nations* (San Francisco, CA: Jossey Bass, 1990).

Potter, S. J., *News and the British World: The Emergence of an Imperial Press System* (Oxford: Oxford University Press, 2003).

—, 'Empire, Cultures and Identities in Nineteenth- and Twentieth-Century Britain', *History Compass*, 5:1 (2007), pp. 51–71.

—, 'Webs, Networks, and Systems: Globalisation and the Mass Media in the Nineteenth- and Twentieth-Century British Empire', *Journal of British Studies*, 46:3 (2007), pp. 621–46.

Pratt, M. L., *Imperial Eyes: Travel Writing and Transculturation* (London: Routledge, 1992).

Preece, W. H., 'The Engineering Conference 1899', *Feilden's Magazine*, 1:1 (1899), pp. 2–15.

Prout, H. G., 'The Economic Conquest of Africa', *Engineering Magazine*, 18:5 (1900), pp. 657–80.

Purkis, A. J. 'The Politics, Capital and Labour of Railway-Building in the Cape Colony 1870–1885' (unpublished PhD thesis, University of Oxford, 1978).

Putnam, R. D., *Bowling Alone: The Collapse and Revival of America Community* (New York: Simon & Schuster, 2001).

Ransome, J. S., *The Engineer in South Africa: A Review of the Industrial Situation in South Africa after the War and a Forecast of the Possibilities of the Country* (London: A. Constable, 1903).

Reader, W. J., *Professional Men: The Rise of Professional Classes in Nineteenth-Century England* (London: Weidenfeld & Nicholson, 1966)

Reynolds, T. S. (ed.), *The Engineer in America: A Historical Anthology from Technology and Culture* (Chicago, IL: Chicago University Press, 1991).

Riffenburgh, B., *The Myth of the Explorer: The Press, Sensationalism, and Geographical Discovery* (Oxford: Oxford University Press, 1994).

Robinson, R. E., 'Introduction: Railway Imperialism', in C. Davis and K. Wilburn, with R. E. Robinson (eds), *Railway Imperialism* (New York: Greenwood Press, 1991), pp. 1–7.

Rolt, L. T. C., *Victorian Engineering* (London: Allen Lane, 1970).

Ross, J. C., 'Irrigation and Agriculture in Egypt', *Scottish Geographical Magazine*, 9 (1893), pp. 170–93.

—, 'Letter to the Editor: The Threatening Destruction of the Temple of Philae', *Academy* (3 February 1894), p. 110.

Rotberg, R. I., *The Founder – Cecil Rhodes and the Pursuit of Power* (London: Southern Book Publishers, 1988).

— (ed.), *Patterns of Social Capital: Stability and Change in Historical Perspective* (Cambridge: Cambridge University Press 2000).

Royal African Society, 'Fifty Years of a British African Society', *African Affairs*, 50:200 (1951), pp. 177–95.

Ryan, J. R., *Picturing Empire: Photography and the Visualization of the British Empire* (Chicago, IL: Chicago University Press, 1998).

Said, E., *Orientalism* (New York: Pantheon Books, 1978).

—, *Culture and Imperialism* (New York: Vintage Books, 1994).

Saunders, C., and I. R. Smith, 'Southern Africa, 1795–1910', in Porter (ed.), *The Oxford History of the British Empire, Vol. III*, pp. 597–624.

Säve-Söderbergh, T., *Temples and Tombs of Ancient Nubia: The International Rescue Campaign at Abu Simbel, Philae and Other Sites* (London: Thames & Hudson, 1987).

Saxenian, A., *Regional Advantage: Culture and Competition in Silicon Valley and Route 128* (Cambridge, MA: Harvard University Press, 1994).

Sayyid-Marsot, A. L. A., 'The British Occupation of Egypt from 1882', in Porter (ed.), *The Oxford History of the British Empire, Vol. III*, pp. 651–64.

Schneer, J., *London 1900: The Imperial Metropolis* (London: Yale University Press, 1999).

Scott, J., *Social Network Analysis. A Handbook: Theory and Analysis* (1991; London: Sage Publications, 2000).

Scott-Moncrieff, C., 'Irrigation in Egypt', *Nineteenth Century*, 17:2 (1885), pp. 343–53.

—, 'The Nile', *Nature* (7 March 1895), pp. 444–6.

Sears, J. E. (ed.), *Who's Who in Engineering 1920–21* (London: Compendium Publishing, 1921).

Shelford, A. E., *The Life of Sir William Shelford, KCMG, Chevalier of the Order of the Crown of Italy, Member of Council of the Institution of Civil Engineers* (privately printed, 1909).

Shelford, F., 'Railway Surveying in Tropical Forest', *PICE*, 133 (1898), pp. 339–50.

—, *Address on West African Railways, the African Trade Section of the Incorporated Chamber of Commerce Liverpool* (Liverpool: Liverpool Chamber of Commerce, 1900).

—, 'To Kumassi by Rail', *Graphic* (22 December 1900), pp. 935–6.

—, 'On West African Railways', *Journal of the Royal African Society*, 1:3 (1902), pp. 339–54.

—, 'Survey of the Niger', *Graphic* (24 June 1907), pp. 1154–5.

—, 'Ten Years of Progress in West Africa', *Journal of the Royal African Society*, 6:24 (1907), pp. 341–9.

—, 'Pioneer Engineering I', *Engineer*, 104 (8 May 1908), pp. 469–71.

—, 'Pioneer Engineering II', *Engineer*, 104 (15 May 1908), pp. 495–6.

—, 'Pioneer Engineering III', *Engineer*, 104 (22 May 1908), pp. 528–9.

—, 'Pioneer Engineering IV', *Engineer*, 104 (29, May 1908), pp. 549–50.

—, *Pioneering* (London: E. & F. N. Spon, 1909).

—, 'Some Features of the West African Government Railways. Including Appendix', *PICE*, 189 (1912), pp. 1–23.

—, 'Sierra Leone in the Making', *Journal of the Royal African Society*, 28:111 (1929), pp. 235–40.

—, 'The Late Mrs J. R. Green and the African Society', *Journal of the Royal African Society*, 28:112 (1929), p. 414.

—, 'Some Notes on the History of the African Society', *Journal of the Royal African Society*, 34:136 (1935), pp. 223–6.

Short, J. R., *Urban Theory: A Critical Assessment* (Hampshire: Palgrave Macmillan, 2006).

Smiles, S., *The Collected Works of Samuel Smiles* (1857–1905; London: Routledge, 1997).

Smith, N. A. F., *The Centenary of the Aswan Dam 1902–2002* (London: Thomas Telford, 2002).

Soja, E. W., *Postmodern Geographies – The Reassertion of Space in Critical Social Theory* (London: Verso Books, 1989).

Stanley, H. M., *The Congo and the Founding of its Free State; a Story of Work and Exploration* (London: Sampson Low, 1885).

Starostina, N., 'Engineering the Empire of Images: Constructing Railways in Asia before the Great War', *Southeast Review of Asian Studies*, 31 (2009), pp. 181–206.

Statham, H. H., 'Philae and the Reservoir', *Builder* (3 March 1894), pp. 165–6.

Stiebing, W. H., *Uncovering the Past: A History of Archaeology* (Oxford: Oxford University Press, 1993).

Stoler, A. L., and F. Cooper, 'Between Metropole and Colony: Rethinking a Research Agenda', in F. Cooper and A. L. Stoler (eds), *Tensions of Empire: Colonial Culture in a Bourgeois World* (Berkeley, CA: University of California Press, 1997), pp. 1–58.

Storey, W. K., *Guns, Race, and Power in Colonial South Africa* (Cambridge: Cambridge University Press 2008).

Sunderland, D., 'The Departmental System of Railway Construction in British West Africa, 1895–1906', *Journal of Transport History*, 23:2 (2002), pp. 87–112.

—, *Managing the British Empire: The Crown Agents, 1833–1914* (Woodbridge: Boydell Press, 2004).

—, *Managing British Colonial and Post-Colonial Development: The Crown Agents 1914–74* (Woodbridge: Boydell Press, 2007).

—, *Social Capital, Trust and the Industrial Revolution 1780–1880* (London: Routledge, 2008).

— (ed.), *Communications in Africa, 1880–1939*, 5 vols (London: Pickering & Chatto, forthcoming 2012).

Taddei, A., 'The London Club in Late Nineteenth Century', *Discussion Paper in Economic and Social History April 1999*, at http://econpapers.repec.org/paper/nufesohwp/_5F028 [accessed 29 September 2009].

Talbot, F. A., *The Railway Conquest of the World* (Philadelphia, PA: J. B. Lippincott, 1911).

—, *Railway Wonders of the World* (London: Cassell, 1917).

Thomas, A., *Rhodes: The Race for Africa* (London: St Martin's Press, 1996).

Thompson, A., *Imperial Britain. The Empire in British Politics c. 1880–1932* (London: Longman, 2000).

—, *The Empire Strikes Back? The Impact of Imperialism on Britain from the mid-Nineteenth Century* (Harlow: Pearson Education, 2007).

Tignor, R. L., 'British Agricultural and Hydraulic Policy in Egypt 1882–1892', *Agricultural History*, 37:2 (1963), pp. 63–74.

—, 'The "Indianisation" of the Egypt Administration under British Rule', *American Historical Review*, 68:3 (1963), pp. 636–61.

—, *Modernisation and British Colonial Rule in Egypt 1882–1914* (Princeton, NJ: Princeton University Press 1966).

Topham, J., 'Scientific Publishing and the Reading of Science in Nineteenth-Century Britain: A Historiographical Survey and Guide to Sources', *Studies in the History and Philosophy of Science*, 31:4 (2000), pp. 559–612.

Tvedt, T., *The River Nile in the Age of the British: Political Ecology and the Quest for Economic Power* (London: I. B. Tauris & Co., 2004).

Warf, B., and S. Aris, *The Spatial Turn: Interdisciplinary Perspectives* (London: Routledge, 2008).

Washbrooke, D., 'Orients and Occidents: Colonial Discourse Theory and the Historiography of British Imperialism', in Winks (ed.), *The Oxford History of the British Empire, Vol. V*, pp. 596–619.

Watson, G., *The Civils: Story of the Institution of Civil Engineers* (London: Thomas Telford House, 1988).

Weiner, M., *English Culture and the Decline of the Industrial Spirit, 1850–1980* (Cambridge: Cambridge University Press, 1982).

Weinthal, L. (ed.), *The Story of the Cape to Cairo Railway and River Route, 1887–1922*, 4 vols in 5 (London: Pioneer Publishing Company, 1924).

Wesselink, H. L., *Divide and Rule: The Partition of Africa 1880–1914* (London: Praeger, 1996).

Whitehouse, B., 'To the Victoria Nyanza by the Uganda Railway', *Journal of the Society of Arts*, 38:2 (1902), pp. 229–41.

Willcocks, W., *Egyptian Irrigation* (London: Millhouse, 1899).

—, *Egypt Fifty Years Hence* (Cairo: Khedival Agricultural Society, 1902).

—, *The Nile Reservoir Dam at Assuan, and After* (London: E & F. N. Spon, 1903).

—, *Sixty Years in the East* (London: William Blackwood & Sons, 1935).

Williams, M. E. W., *The Precision Makers: A History of the Instruments Industry in Britain and France 1870–1939* (London: Thomson Learning, 1994).

Wills, W. H., *Anglo-African Who's Who and Biographical Sketch Book* (London: Upcott Gill, 1907).

Wilson, E., *Our Egyptian Obelisk: Cleopatra's Needle* (London: Brain & Co., 1877).

Wilson, J. F., and A. Popp (eds), *Industrial Clusters and Regional Business Networks in England 1750–1970* (Oxford: Oxford University Press, 2003).

Winks, R. (ed.), *The Oxford History of the British Empire, Vol. V: Historiography* (Oxford: Oxford University Press, 1999).

Wolfe Barry, J., 'Presidential Address', *PICE*, 134 (1898), pp. 1–13.

Wood, C., *Olympian Dreamers: Victorian Classical Painters* (London: Constable, 1983).

Woodrow, H., *Tales of Victoria Street: The Story of the Association of Consulting Engineers* (London: Association of Consulting Engineers, 2003).

Zimmer, G. F., *Engineering of Antiquity and Technical Progress in Arts and Crafts* (London: Probsthain & Co., 1913).

INDEX

Academy, 51
accountants, 37, 41
Admiralty, 39
advertisements, 16, 17, 26, 29, 31, 132–4,
    174n35, 176n76, 196n106
Africa
    Cape–Cairo Railway, 20, 23, 118–19,
        131
    ICE advisory committees in, 100, 162
    ICE members in, 91, 93, 94, 96, 97, 98
    sensationalizing, 118, 122, 123–4, 136
    see also countries
African and Eastern Engineering, 30
African Engineering, 15, 18, 26–30, 46,
    80–1, 125–6, 162
African Society (later Royal African Soci-
    ety), 13, 126, 128
African World, 27, 126
Allen, Charles Edgar, 21, 25
Andes, 48, 64
Appleby, Messrs, 62
apprenticeships, 58, 59, 60, 62, 63, 65, 67–8,
    73
Arapostathis, S., 135–6
Argentina, 52, 53, 64, 66, 190n72
Ashanti Goldfields Corporation, 126
Association of Consulting Engineers,
    172n25, 196n106
Aswan Dam, 2, 20, 40, 62, 75, 85, 142–59,
    201n87
Athenaeum, 151
Athenaeum Club, Pall Mall, 42, 157
Australia, 46, 65
    ICE advisory committees in, 100
    ICE members in, 91, 96, 97, 100

New South Wales, 39, 52, 53, 60–1, 97,
    100, 164
    railway bridges, 66
    railways, 53

Bacon, Francis, 21
Baker, Sir Benjamin, 42, 85, 87, 95, 96, 105,
    107, 128, 129–30, 164, 183n32
    Aswan Dam, 137, 168
    commemorative window in Westminster
        Abbey, 167–8
    consulting engineer to Aswan Dam, 154
    Forth Railway Bridge, 143, 168
    and John Fowler, 60, 62–3, 64, 69, 74,
        75, 130–1
    freshwater canal proposal, 130–1
    obituary, 184n41
    Philae international commission, 143–4,
        147, 151–2, 156–7, 158
    presidency of ICE, 70
    transportation of Cleopatra's Needle,
        141, 142
Baker, Sir Samuel, 122
Baker & Hurtzig partnership, 63
Ball, Michael, and David Sunderland, 37, 40
Ball, W. Valentine, 6, 132–3
Barbados, 63
Barlow, John Henry, 73
Barlow, William H., 42
Baro–Kano railway, 82–4
Barrett, John, 23
Barton, Ruth, 42
Bechuanaland, 40, 75–6
Becket, Bernard H., *Scientific London*, 89
Bell, Horace, 97
Benjamin Baker Medal, 89
Bennett, William, 100